THE BEST OF ALL
POSSIBLE WORLDS

The Best of All

Possible Worlds

MATHEMATICS AND DESTINY

Ivar Ekeland

The University of
Chicago Press
Chicago and London

IVAR EKELAND is professor of mathematics at the University of British Columbia and director of the Pacific Institute for Mathematical Sciences. He is the author of *Mathematics and the Unexpected* and *The Broken Dice* (University of Chicago Press 1988 and 1993).

The University of Chicago Press, Chicago 60637
The University of Chicago Press, Ltd., London
© 2006 by The University of Chicago
All rights reserved. Published 2006
Printed in the United States of America
15 14 13 12 11 10 09 08 07 06   1 2 3 4 5
ISBN-13: 978-0-226-19994-8 (cloth)
ISBN-10: 0-226-19994-0 (cloth)

Originally published in French as *Le meilleur des mondes possibles: Mathématiques et destinée* © Editions du Seuil, 2000.

Library of Congress Cataloging-in-Publication Data

Ekeland, I. (Ivar), 1944–
    The best of all possible worlds : mathematics and destiny / Ivar Ekeland.
        p.   cm.
    Includes bibliographical references and index.
    ISBN 0-226-19994-0 (cloth : alk. paper)
        1. Science—Mathematics. 2. Mathematical analysis. 3. Logic, Symbolic and mathematical. 4. Human behavior. 5. Ethics. I. Title.
    Q172.E36   2006
    509—dc22
                                                                    2006001984

∞ The paper used in this publication meets the minimum requirements of the American National Standard for Information Sciences— Permanence of Paper for Printed Library Materials, ANSI Z39.48–1992.

(CONTENTS)

Introduction, 1

1 Keeping the Beat, 3

2 The Birth of Modern Science, 24

3 The Least Action Principle, 44

4 From Computations to Geometry, 79

5 Poincaré and Beyond, 102

6 Pandora's Box, 117

7 May the Best One Win, 129

8 The End of Nature, 145

9 The Common Good, 166

10 A Personal Conclusion, 182

*Appendix 1: Finding the Small Diameter
of a Convex Table, 193*

*Appendix 2: The Stationary Action Principle
for General Systems, 195*

Bibliographical Notes, 197

Index, 199

THE OPTIMIST BELIEVES that this is the best of all possible worlds, and the pessimist fears that this might be the case. Life is not always kind to humans, and from the earliest times on, they have wondered why, and turned to priests or philosophers for an answer. There was, however, a window of two centuries, between 1600 and 1800, during which some scientists felt that they could contribute to the answer. Chief among them was Maupertuis, a French polymath who was at once an explorer, a scientist, a philosopher, and a courtier. He discovered that all the laws of physics were mathematical consequences of a single idea, which he called the least action principle: everything happened as if a certain quantity, called the action, was to be made as small as possible. If one accepted that idea, then all the laws of physics could be derived by mathematical techniques. He then crossed the bridge between science and metaphysics by stating that similar principles were at work in all of creation, so that, for instance, God ordained the course of history so that the total amount of suffering incurred by humankind should be minimal. This started a huge controversy, and Maupertuis was ridiculed by Voltaire in his famous novel *Candide*, later a musical by Leonard Bernstein, as the philosopher Pangloss, who walks through an ever-worsening succession of disasters while blindly claiming that all is well that ends well in the best of all possible worlds.

Maupertuis deserves a better fate. Scientifically speaking, his least action principle is basically sound. It has been taken over, transformed (perhaps beyond recognition), and improved upon continuously, and recently it has led to a series of breakthroughs in mathematics. I have had the privilege of being involved in this research, which is all the more fascinating since it has its roots so far back in history, and I want to share some of my experience and enthusiasm with others. In addition, Maupertuis may have been the first person to understand how important the idea of optimizing—of devising systems which would function in the best possible way according to some criterion—would become in the modern world. I have tried to follow this idea as it moved from physics into biology, and then into the social sciences. This itinerary more or less coincides with my personal history. I have moved from mathematics to mechanics, and then to economics, always following the trail of optimization from one field of knowledge to the other. During this journey, my scientific interests have shifted accordingly, and I now find myself studying human behavior. The more important questions appear later in my scientific career, as if I needed the accumulated knowledge and experience to finally find out what the right questions are.

What are humans? What are we trying to do to ourselves and to the environment? This is no longer a philosophical question. The way we are using up the resources of the planet, and fighting among ourselves in the process, is now turning it into an immediate and practical issue. This book tries to show how this question slowly emerged from the progress of science, and points in some directions for the future.

(CHAPTER 1) **Keeping the Beat**

"BEFORE WE PROCEED, we must be aware that every pendulum keeps its beat so well defined and fixed that it is not possible to have it move according to any other period than the only natural one." Thus speaks Galileo in his *Discourses and Mathematical Proofs Concerning Two New Sciences*, the last book he ever published (1638). He died four years later, leaving a rich scientific legacy, of which this simple statement may be the most important part: as a fact, it quickly turned out to be false, but it changed our ideas about physical motion, and inspired a new technology to measure time.

A pendulum is simply a small load suspended to a string or to a rod fixed at one end. If left alone it ends up hanging vertically, and if we push it away from the vertical, it starts beating. Galileo found that all beats last the same time, called the *period*, which depends on the length of the pendulum, but not on the amplitude of the beats or on the weight of the load. It also states that the period varies as the square root of the length: to double its period, one should make the pendulum four times as long. Making it heavier, or pushing it farther away from the vertical, has no effect. This property is known as *isochrony*, and it is the main reason why we are able to measure time with accuracy.

It has been said that Galileo discovered that law during a service in the Pisa cathedral, by comparing the oscillations of the great lantern hanging in the nave to the beats of his own pulse. What a beautiful symbol! The great cosmic cycles, the succession of night and day, the phases of the moon, the coming of the tide, the return of the seasons, have always been the background against which history is staged. But there is also for each of us a smaller companion, which measures not cosmic, but biological, even personal, time: our own pulse is a natural wristwatch. Comparing the rhythm of nature to that of our blood gives us the idea of a standard time, which should be both universal, valid for everyone at once, like the oscillations of the lantern, and homogeneous, reproducing itself regularly, like the beats of our heart. It is a truly revolutionary idea, contrary to all the experience which humankind has gathered since the earliest times: all natural rhythms are variable and irregular. The pulse beat is not the same across individuals, and it is affected by emotional or physical strains. Daylight varies according to the latitude and the season, the lunar month changes as well, and defining correctly the year is a major astronomical problem. Coordinating all these rhythms, to keep Christmas in midwinter for instance, required the invention of the Gregorian calendar with its complicated rules about leap years. It still is not good enough, because these rhythms change: the rotation of the Earth is slowing down, so that the day lengthens slightly, and once in a while the atomic clocks that keep standard time have to be pushed forward one second.

Today hours are constant: an hour is an hour, anywhere in the world, at any time, just as a meter is a meter and a pound is a pound. But this is quite a modern idea: for our ancestors, hours were uneven. In classical antiquity, there were twelve hours between sunrise and sunset, and twelve hours between sunset and sunrise. So day hours and night hours (aka vigils) had different durations, except at the spring and fall equinoxes. Coming to work in the fields at the eleventh hour meant that most of the day had gone by; little wonder that those who had been around since dawn found it unfair to be paid no more than the latecomer, as in the evangelical parable. The duration of hours varied with season and location: summer hours were different from winter hours, Florence hours were different from Rome hours (not that there was any direct way to compare them in those times).

Not so with the pendulum beat. The great lantern of the Pisa cathedral beats for all to see, and each and every one of its oscillations has the same duration. They slowly dampen, and eventually the pendulum will stop, but a whiff of air or a pull on the ropes will start it again, always with the same period, measuring out equal intervals of time. Bring it to Rome, and it will keep the same beat as in Pisa, by day as by night, in summer as in winter. This is Galileo's great discovery: the pendulum provides us with a natural way to measure time in a universal and homogeneous way. It divides time into intervals of constant duration, unlike the day, the month, or the year, which are difficult to carry around and vary according to place and date.

Between the fourth century BC, and the fourth century AD, over a period stretching eight hundred years, there flourished in Alexandria an extraordinary school of Greek mathematicians, starting with Euclid, the legendary founder of geometry, and ending with Hypatia, probably the first woman to leave a name in mathematics. Their work was familiar to Galileo and to all the scientists of his time: they had explored essentially all the possibilities offered by the ruler and the compass, and no better instruments were available. The basic shapes of geometry still were those which could be constructed using only ruler and compass: lines and circles, of course, but also the three conics, ellipse, parabola, and hyperbola, on which no progress had been made since the comprehensive treatise of Apollonius, written in Alexandria in the course of the third century BC. At about the same time, another great scientist, Archimedes, showed how to compute the areas of these curves, and also the volume of the bodies they generate by rotating around an axis. The technology in Alexandria was impressive as well, probably better than that which was available to Galileo. Many treatises on architecture and engineering have survived, and the renown of some of their realizations have crossed the bridge of centuries. The war machines Archimedes built kept at bay for three years the Roman army besieging Syracuse; the great beacon of Alexandria harbor could be seen from thirty miles at sea.

Galileo does to time what the great geometers of antiquity did to space: he turns it into a homogeneous and measurable quantity. Whereas the Greeks had a well-established theory of space, which remained fruitful and essentially unchanged until the non-Euclidian geometries were discovered in the nineteenth century, they did not have a corresponding theory for time. They had a grasp of statics, not of dynamics. Every kind of motion, be it of an arrow flying toward its aim, a runner catching up with a tortoise, or a stone thrown in the air, was a problem for them. What force is driving the stone, once it has left the thrower's hand? How can the runner catch up with the tortoise? Mark the position of the tortoise and wait till the runner has reached it; but the tortoise has progressed in the meantime, so there is a new position to be marked, and an additional time to wait before the runner reaches it; but the tortoise has moved again, so that it will always be slightly ahead, and the runner should never catch up. This is Zeno's paradox, clearly a question posed by someone who has a better grasp of space than of time.

Unlike geometers, the Greek physicists did not worry about the possibility of motion—they just took it as a fact—but they looked for its causes. The most influential work on the subject was the *Physics* of Aristotle, written in the fourth century BC. This was the main influence that Galileo would have to fight to establish the "new science," as he called it. Aristotle's physics is plain good sense: whenever something moves, something else must be driving it, and whenever the driving force stops, the driven object must stop. It is not without its problems: how come the stone I throw does not fall directly to the ground as it leaves my hand? Why does it rise first, and then fall back? It is quite interesting to see trajectories of projectiles drawn in Galileo's times: the projectile is shown to rise in an arc, and then to fall steeply, almost vertically, as if it had been dropped from the high point of the trajectory. This is in accordance with Aristotle's teaching, but it is not what actually happens: the second part of the trajectory is an arc, symmetric to the first one. It was clear enough why the stone would eventually fall, but much ingenuity was spent in explaining why it would have to rise first, and the air was suspected to play a role in carrying it. In short, the Greeks did not develop a theory of time and motion analogous to the theory they developed for space and shape.

No wonder: in Greek philosophy, motion is synonymous with change, and hence with imperfection. Something truly perfect would not change, it would neither grow nor decay, it would be unalterable and eternal. In Platonic philosophy, perfect objects do exist; they constitute the only true reality: what we see during our lives are only poor reflections of these ideal objects, mere shadows on a wall. After our death, however, we will be allowed to contemplate the originals, to see good everlasting, truth everlasting, beauty everlasting, and we will carry some memory of them in our later lives. We do not discover mathematical truths; we remember them from our passages through this world outside our own. There is a famous scene in the dialogue *Meno*, where Socrates leads an uneducated slave into "remembering" that the diagonal c of a rectangle is related to its sides a and b by the famous theorem of Pythagoras: the square of c is equal to the sum of the squares of a and b. It is well worth reading, and is an example of good teaching. Socrates never tells the slave anything; he simply asks him the right questions, in the right order, and lets him grope around until he suddenly sees the theorem, sees it as truly and self-evidently as if he had always known it. In fact, says Plato, Meno knew that theorem because he had already seen it, in the world of eternal truths which his soul had visited before it was sent back to Earth in the body of a slave. Socrates was wont to say that he held the same profession as his mother, a midwife, because he delivered souls of the burdens they did not know they carried, just as she delivered pregnant women of their unseen progeny.

In the Platonic tradition, truth is never discovered; it is remembered. Between two successive lives, the soul journeys through the realm of the dead and the unborn, to contemplate one more time the Ideas, perfect, immutable, and eternal, which are the blueprints of everything it will meet during its travels on Earth. Even the word "theory" is a witness to that conception of knowledge: in Greek, *theorein* means "to see," and *theoreia* means "the things which have been seen." Anything transient, such as physical motion, has no place among the Ideas; we can have no "theory" for it because we cannot have seen it before our lifetime. Only immobility can carry some kinship with the perfection of Ideas, and there is indeed in Greek physics a well-developed theory of rest, or *equilibrium*, in more scientific terms: the most famous instance is the theory of equilibrium of fluids, which allegedly sent Archimedes running naked through the streets of Syracuse in the first joy of discovery.

If an object is left at equilibrium, on its own, it will stay there forever. To drive it away from equilibrium, we must exert some force on it, preferably by direct contact; this force is the cause of the motion, and as soon as the cause disappears, the motion should stop. That is the intellectual framework in which Aristotle and his successors try to understand the various kinds of motions that surround us in the real world. This is not without difficulties; to explain the motion of the stars, for instance, which they imagine as luminous dots pinned on a gigantic sphere surrounding us and on which the Sun travels daily, they have to call in legions of angels or demons which push the exterior of the heavenly sphere to make it rotate. During antiquity and the Middle Ages, the world is seen as full of a bewildering variety of motions, of objects scrambling back to equilibrium. There is no general theory; for each motion, some reason must be found why that particular object should have fallen out of equilibrium at that particular time, and how it will proceed to reach anew a state of rest. This is not an easy task, and some answers had evaded scientists for centuries.

For instance, since Roman times it had been noticed that water could be pumped no more than ten meters high at one time; if higher reaches were sought, more pumps were needed, each one pumping water into a basin for the next one to pump from, but each pump could do no better than ten meters. The explanation given was that nature had some kind of distaste for vacuum, and would therefore tend to fill every empty space in the universe before reaching an equilibrium. Why this particular distaste would stop at ten meters, or why the universe would be content with filling the pumps up to ten meters' height, was beyond the reach of the most imaginative explanations. In short, until Galileo, physical motions are seen as perturbing the fundamental order of the universe, which is mirrored by classical geometry. Motion is disorder. The natural state of a physical object is to be immobile.

That day in the Pisa Duomo, Galileo sees the opposite: back and forth swings the great lantern, back and forth. It goes through the vertical position, runs up the other side, hesitates a moment, and then swings back. Eventually it will slow down; its swing will gradually wind down, keeping the same beat, until it finally hangs motionless, the smoke from the candles rising vertically to the gilded ceiling. Why should this position be more natural than this symmetric motion, back and forth, back and forth, with a majestic regularity? What is there to prevent it going on forever? Is it slowing down of itself, or are we witnessing the effect of friction, exerted by the surrounding air and the suspending ropes? Would these not count as imperfections, against the perfection of an oscillatory motion, indefinitely going through the same positions at regular intervals? Certainly the air does not sustain the motion, as we can see from the trailing smoke of the candles: it must be that the motion sustains itself, and it is slowed down by its surroundings. If these could be corrected, the pendulum would beat forever, like the pulse of this great cathedral. And it would spin out, forever and ever, intervals of equal duration, which could thenceforth be used to measure time, just like folding rules are used to measure lengths.

Galileo's theory of the pendulum—and we may use that word as the ancient Greeks did, equating theory to vision, because Galileo actually saw it that day in the cathedral, and all his subsequent work was to remember and understand what he had seen—consists first in the basic intuition that motion need not be from one equilibrium to another, that a pendulum will swing forever, pausing briefly twice a beat as it reaches the top of its trajectory before falling back again. If it eventually slows down and stops, it is because of various imperfections which have to be taken into account, the correction of which will lead, if not to perpetual motion, at least to prolonged life. The second great idea was that all oscillations of the same pendulum, large or small, have the same duration, depending only on its length (this is isochrony, as I mentioned earlier). For the first time in history, humankind had found a chronometer, an instrument which measured time with accuracy and was easy to carry about. Two pendulums of equal length, one in Paris and one in Rome, would have the same beat, regardless of the amplitude of their swing. A piece of string ten inches long is a simple chronometer. Just attach a weight to one end and let it swing from the other. One full beat lasts almost one second; there are sixty beats to the minute, and, if you are patient enough, 3,600 to the hour. A pendulum which is four times as long will be twice as slow: a string of one meter will have a half beat of one second.

This was a remarkable connection between geometry and dynamics. Shortly after, mathematicians would conquer time as they had already conquered space. The isochrony of the pendulum is not factually true; it is an idealization, as the straight lines and the circles we learn about in geometry are idealizations of what we actually draw on sand or on paper. An actual pendulum will beat more slowly as its swings become wider, as we can convince ourselves by setting two pendulums of equal length side by side and starting them from different positions. The duration of the beat—the *period*—increases with the amplitude; small oscillations, the ones which remain close to the vertical, have smaller periods than large ones. However, the discrepancy is very small as long as the oscillations are kept small. The influence of the amplitude on the period starts making itself felt only for large deviations from the vertical. Of course, discrepancies which are very small in themselves add up in the course of a day or a week, and the only safe solution is to keep the pendulum swinging with exactly the same width, which is precisely what the mechanism of grandfather clocks is designed to do, and why they have to be wound up in the first place. But Galileo's idea is right, just as the idea that a straight line is infinite in both directions and has zero thickness is right. We all know that we cannot draw the line farther than the page allows, and we do not need a magnifying glass to know that it is as thick as the lead of the pencil we have drawn it with. But we understand the idea, and it is useful to us in building bridges and roads, and in drawing boundaries. Similarly, it is only for small amplitudes that pendulums behave according to Galileo's prescription, but it is a good starting point for us to understand more general oscillations and to build timepieces.

This is the true Galilean revolution. It is told that, after kneeling down in front of the tribunal to forswear the Copernican view that the Earth moves around the Sun, Galileo touched the ground while standing up and said, "And yet, it moves!" He was speaking of the Earth, of course, but it may be said as fittingly of the pendulum, a humble object which his genius had turned into a mathematical idea, as sharp and fruitful as the idea of a circle. The idea of periodic motion was the missing link between space and time. No longer could it be said that motion is ephemeral and transitory, a simple shift from one equilibrium to another: Galileo's pendulum moves unchanging. There is no cause to its motion; it has no beginning and no end. Actual time, as we experience it, is bounded at both ends, by our birth and our death, or, if we probe deeper, by the birth and death of the universe. Not so, Galilean time, since his ideal pendulums keep beating forever. In this respect time is very similar to geometrical space, as defined by the great Alexandrines since Euclid: they understood it as unbounded, although physical space certainly is bounded, either by the limits of the Earth, or by the heavenly sphere which surrounds it. Put in the middle of this unbounded space a Galilean pendulum, beating time like the lantern in the Pisa cathedral, and you have the modern universe, the framework of science to this day.

Galileo's idea also provides us with a natural unit to measure time by. As we saw, neither the year nor the day is satisfactory, for they vary with the position and the date; besides, these are large intervals of time, and it is not obvious how to measure smaller ones. But choose a particular pendulum, a pendulum ten inches long, for instance, and define the second to be its period, that is, the duration of one full beat. Defining the standard unit of time in this manner is very similar to the way we used to define units of length; during the French Revolution, for instance, one meter was defined to be the distance between two notches on a certain rod made of an alloy of platinum and iridium. This precious rod was solemnly buried, together with two copies, on September 28, 1889, in a vault at the Breteuil Observatory near Paris, together with the standard kilogram and six copies. More copies were carefully made, checked against the original, and sent to other places to make more copies, right down to schoolchildren's rulers. We could imagine the standard unit of time being defined in a similar way, as the half period of a pendulum one meter long for instance, which would be buried along with the standard units of length and weight, and somehow kept moving. This is not a practical definition, for in fact the period of a pendulum depends on the strength of the gravitational pull, and that changes with the geographical position, because the Earth is not a perfect sphere. Perfect copies of the standard pendulum would have different periods in different places. But let us pursue Galileo's dream a little bit further.

The problem of measurement is not completely solved by defining a unit. We must also show how to divide it into subunits. For units of length, this problem is solved by one of the earliest results of Greek geometry, a theorem attributed to Thales from Miletus, who is said to have predicted an eclipse of the Sun which occurred in 585 BC. It may well be that Thales benefited from the results of Babylonian and Egyptian science, and indeed this theorem is so fundamental to measurement that it must have been known much earlier. Essentially, it says that if you are able to multiply your unit by ten (that is, to replicate it ten times and place the resulting copies end to end on a straight line), then you are also able to divide it by ten. Of course, there is nothing special about the number ten in this result, and the same holds for other numbers, if you do not fancy the metric system. But for units of time, there is no theorem of this kind: an hour is sixty minutes, to be sure, and if you can count minutes you can count sixty of them to make an hour, but this will not help you toward timing events that last less than one minute, like a hundred-meter dash. Measuring time with a pendulum provides an easy answer: if you want your pendulum to beat ten times faster, make it hundred times shorter. If a pendulum of one meter has a half beat of one second, a pendulum of one centimeter will have a half beat of one-tenth of a second. Such a pendulum may be hard to build, and harder to keep beating, but this is the right idea; send scientists and watchmakers to work on it, and in a couple of centuries you will end up with the ultraprecise timepieces which adorn our wrists.

Precision was something new to timekeeping. In geometry, it came as a matter of course. Archimedes, for instance, had written an essay *Measuring the Circle*, devoted to finding the exact value of the ratio *P/D*, where *P* denotes the circumference of the circle and *D* its diameter. This is the famous number π; Archimedes proves that it lies somewhere between 223/71 and 221/70, and gives a numerical procedure to compute it to any desired accuracy. Archimedes' procedure was perfected through the years, and in 1593 the French geometer Viète knew the first seven decimals $\pi = 3.1415926$. Today better procedures and automated computations have yielded billions of digits of π; in fact, we can now compute directly any given digit without bothering with the intervening ones. My point here is that our knowledge of π is so precise that it has long since ceased to have any physical relevance. Already in Galileo's time, to tell the difference between 3,1415926 and 3,1415927 would require that one have instruments to build circles and measure lengths within an accuracy of one part in one billion, way beyond the technical capacities of that time. There is no hope of ever checking experimentally the digits of π beyond the first ten or so. The number π itself, however, exists for mathematical, not physical, reasons, and has infinitely many digits, of which we know only the first fifty billion or so: there is no limit to precision in mathematics. After Galileo, the same principle will apply to chronometry. There is no more problem in describing very small durations, one thousandth of a second, say: it is simply the half beat of a pendulum one thousandth of a millimeter long. Such a pendulum may be difficult to build, and even more to observe, but it raises no theoretical difficulty; it is an ideal object, as real as the thousandth digit of π. Henceforth, in dynamics as in geometry, precision is mathematical, that is to say infinite, and we will be able to carry computations as far as we wish.

With this new degree of precision in measurement, new problems can be raised. To measure a length, one basically has to establish a coincidence, that is, to bring two objects to the same location: both endpoints of the length to be measured must coincide with some graduation on a ruler. To measure a duration, one must establish a simultaneity, that is, to have two events occur at the same instant: the runner starts just as the pendulum is at the top of its swing, and crosses the end line just as it is at the top of another swing. But what does it mean that two events occur "at the same instant"? If both occur at the same place, or close by each other, the meaning is clear enough, but what if they happen far apart? So far apart, for instance, that they cannot be observed together? Galileo, kneeling in the Pisa cathedral, may well count the oscillations of the great lantern against the beat of his own pulse. But does it make sense to ask what is happening in China at the same time? Does simultaneity bear traveling? Can one imagine a slice of time through the universe, the lantern in Pisa frozen in its swing, the emperor in midstep, the planets in their orbits, the galaxies in their swirl, all caught at the very same instant? The whole history of the universe would then be but a succession of such slices, as a motion picture is a succession of photographs.

There would be no problem if, for instance, light propagated instantaneously: then events observed from afar would occur at the very moment they are seen, and simultaneity would be easy to establish. But this is not the case, and then one needs to take into account the distance from the observer, the path of light and the speed of propagation. In other words, simultaneity cannot be established directly, like coincidence: a full-fledged theory of light is needed just to state that two events occurred at the same time (unless they occurred at the same place). For instance, if the universe is imbedded in the three-dimensional infinite space of Greek geometry, and if light propagates at a constant speed along straight lines, 300,000 km/s, say, then any event I observe now from a distance of 300,000 kilometers must have occurred one second ago. This is the kind of theory Galileo had in mind, the consequence of which is that, indeed, there is a global and universal meaning to simultaneity. An observer from Sirius, if he had a telescope sharp enough to see through interstellar space, the Earth's atmosphere, and the dome of the Pisa cathedral, could watch the oscillations of the great lantern; they would have precisely the same duration for him as for the congregation, so that, for instance, they could serve to define a common unit of time on Sirius and on Earth. He would know also that the oscillations he was witnessing had taken place on Earth about 8.6 years before, and were therefore simultaneous with certain events in Sirius he could name. Piecing together such observations, spread over a long period of time, one could arrive at a picture of a portion of the universe as it was one thousand years ago, or one million years ago—the longer the delay, the larger the region covered.

So the idea of the whole universe at a given instant does make sense, and it certainly was in Galileo's mind, as it is in ours. One can imagine, for instance, every astronaut in every galaxy carrying a clock, and each clock showing the same date and time—UST, Universe Standard Time. Whenever two astronauts meet, and wherever they come from, their watches would show the same UST. A traveler leaving Earth in a space-ship will find on his return that he has aged precisely as much as those who have stayed home, and the watch he is carrying would show exactly the same time as the one he has left behind.

This, of course, is in sharp contrast with the modern theory of light, due to Albert Einstein, called special relativity: the space traveler will find on his return that the watch he is carrying is slow with respect to the watch he has left, and that more time has elapsed for the people on Earth than for him. Clearly, one cannot define any UST in this theory, and neither can one decide whether two events which happened at different places are simultaneous or not. How could the space traveler reconcile his calendar with the one on Earth? He left two years ago (his time), or twenty years ago (Earth time); so far, so good. The traveler and the people who stayed can agree on that, because they were together as the trip began, and as it ends, they can simply compare their watches. But suppose the traveler is told his mother died three years ago; does it make sense for him to be asking himself, "What was I doing when it happened?" In fact, within the theory of relativity, it does not make sense: simultaneity can be established only for events which occur at the same location. There is no way to carry time from one place to another. Suppose I set my watch to some clock, which is supposed to deliver UST, and then travel to another place where I set a second clock to my watch. If I then go back to the first clock, I will find it no longer agrees with my watch! Can I claim that the second clock still shows UST? The discrepancy is extremely small (in fact, cannot be detected) as long as the various velocities involved stay safely away from the speed of light. But it becomes significant (and has to be taken into account) as soon as this speed is approached, which happens quite often at the subatomic level.

Galileo's theory and Einstein's hold true at different scales. Certainly, science did not require anything beyond Galileo's theory of space and time until it started investigating electromagnetic waves, including light, toward the end of the nineteenth century. Until then, the idea of some Universe Standard Time was a perfectly sound one, and it is still good enough for most of today's science, as long as one keeps away from very large scales (cosmology) or very small ones (subatomic particles).

Constructing a timepiece that would keep Paris (or Greenwich) time while being carried around the world, sometimes in less than comfortable conditions, very quickly became a major technical challenge. It was not a question of putting Galileo's ideas to the test, but of determining the position of ships at sea. Two numbers were needed, latitude and longitude. The former was determined by measuring the maximum height of a star, or of the Sun, above the horizon, and comparing it with astronomical tables which give this height as a function of the latitude and the date. It was not an easy matter, it required precise sighting instruments and reliable numerical tables, but the appropriate techniques had been inherited from classical antiquity and developed by the Arabs. Determining the longitude, on the other hand, was an unsolved problem. Theoretically, it was easy enough: just clock the precise time when the Sun reaches it peak. This is noon, at the spot where you are; if you happen to know the time it is in Paris (or Greenwich) at that time, then the difference will tell you precisely how far away you are from the Paris (or Greenwich) meridian; that is, you will have the longitude.

Before radio was invented, the only way to know Paris time was to carry it around with you, in the hope that your watch was neither fast nor slow. In July 1714, the British Parliament offered a prize of £20,000 for a method of determining longitude to an accuracy of half a degree— sixty nautical miles. This amounted to building a chronometer which would not lose or gain more than two minutes. The prize was finally won by John Harrison, whose marine chronometer H4 crossed the Atlantic in 1762, having lost less then five seconds in eighty-one days at sea. Just imagine the conditions on board a ship in those times, the perpetual rocking motion, the violent shock when a wave broke against the hull or when a gale filled the sails, the changes in temperature, humidity, and pressure between the Thames and the American coast. Harrison's chronometer was a miracle of precision: keeping time within five seconds meant measuring longitudes within 1.15 nautical mile. It meant being able to draw accurate charts of the world, to locate precisely dangerous coasts and isolated islands, and it was well worth the prize, which was finally, and not without trouble, awarded to Harrison. But it was also a confirmation of Galileo's theory of universal, or absolute, time: if you carry a perfect watch around the world, it will still be on time when it gets back.

Let us now investigate how Galileo's theory of the pendulum was used to devise accurate watches, as we know them today. This will help us gain a better understanding of Aristotelian physics, and of the revolution Galileo brought about.

The earliest timepieces, of course, were the various kinds of sundials that projected the trajectory of the Sun during the day: the position of the shadow gave the hour, and its length gave the date. They were unsuitable for measuring anything smaller than a quarter of an hour, and in addition they did not work if the Sun was hidden. So, during antiquity and the Middle Ages, other instruments were developed, based on the idea of measuring the length of time which elapses as some system goes from one state to another, just as Aristotelian physics would suggest. It could be sand, or water, flowing from an upper chamber to another: this gives an hourglass, or a clepsydra. It could also be a set of weights falling down: this is the principle of weight-driven clocks. All these timepieces made use of transitory motions, which have a beginning and an end, and which have to be restarted by human intervention when they have stopped: the clepsydra must be refilled, the hourglass must be turned over, the weights must be drawn up. This is not to say that these were rudimentary instruments: on the contrary, they were often built with great ingenuity, and many technological problems had to be solved along the way. In a clepsydra, for instance, the level of water in the upper chamber must not be allowed to drop; otherwise the downward flow would slow down, and the clepsydra would keep uneven time. As these timepieces were perfected, more and more was required of them, and toward the end of the Middle Ages, clocks were built which showed the date, the phases of the Moon, and the position of the Sun in the constellations.

However, all these instruments remained imprecise. Flowing water or falling weights will not divide time in perfectly equal intervals. If you have a pendulum, it is easy to measure out equal time intervals: just count the beats. But if you have a clepsydra, or an hourglass, performing the same task becomes much more difficult: you would have to ascertain that the same amount of water has flowed down, or that the weights have fallen an equal height. This is the kind of problem that Galileo met when he performed his famous experiments on falling bodies. He used a clepsydra, and since it was impossible to measure small intervals of time, he slowed down the falling motion by the simple device of having spherical balls roll down a gentle slope instead of dropping them from a height. In his own words, "As for measuring time, we did it with a large bucket of water, which we hung up at a certain height, and from which a trickle of water flowed by a small tube welded to the bottom into a small glass, all the time the ball was moving. The quantities of water were then weighed on very precise scales, and the differences and proportions of weights yielded the differences and proportions of times."[1]

One can easily imagine the precision of such measurements, especially since the motions, even after being so ingeniously slowed down, lasted no more than a few seconds. In fact, in Galileo's work, the new laws of physics are supported more often by mathematical or philosophical arguments than by experiments, the latter being so imprecise that they are either inconclusive or in contradiction with the theory. For instance, in a letter dated March 13, 1641, Vincenzo Renieri informs Galileo that he has climbed the famous leaning tower in Pisa, that he dropped from the top two balls of the same size, one in wood and the other in lead, and that the second one fell faster: upon its arrival, the wooden ball was still almost three yards from the ground. Between 1640 and 1650, Giambattista Riccioli performed in Bologna a series of experiments, and found that two clay balls with the same size, one weighing ten ounces and the other twenty, dropped from 312 feet, did not reach the ground at the same time: there was a fifteen-foot difference between them. All this, of course, was in seeming contradiction with Galileo's laws, which asserted (among other properties) that, in a vacuum, all heavy bodies fall with the same speed. The differences pointed out by Renieri or Riccioli were due to air friction, and would certainly have seemed quite troubling at a time when the existence of air as an independent medium was far from obvious.

Galileo got in trouble with the pendulum as well. Today, it is clear that his assertion that all beats should have the same duration, regardless of their amplitude, is just wrong. In fact, the duration of the beats increases with the amplitude, and it is quite easy to check. Take a rod, suspend it from one end, and you have yourself quite a decent pendulum. Now push the other end farther and farther away from the vertical, even above the horizontal, giving the pendulum greater and greater amplitude; the duration of the beats will noticeably increase. In fact, for the greatest possible amplitude, when the rod is upside down, the rod will be in equilibrium, so that the pendulum does not beat at all; a mathematician will see it as a beat with infinite duration. This equilibrium is unstable: touch the rod ever so slightly and it will start falling, slowly at the start, then faster and faster; by arranging the initial fall to be very slow, we can get as slow a beat as we wish. So the truth of the matter is that the period of the pendulum, that is the duration of the beats, increases to infinity as the initial position approaches the upright one. Already in 1644 Mersenne had pointed out

1. *Discorsi* (Leyden: Elsevier, 1638), second day.

that the period increases with the amplitude. He had done it by the simple device of constructing two identical pendulums and starting them simultaneously from different angles: it was immediately apparent that the wider angle gave a slower beat. The difference, which is trifling as long as both amplitudes are small, becomes significant when they are large. However, Mersenne also checked that the period did not depend on the material or the weight of the pendulum, and that it was proportional to the square root of its length, in accordance with Galileo's laws.

And yet, Galileo clung to the isochrony as if it were an experimental truth. He saw it as a confirmation of the laws he had discovered for the motion of falling bodies. In the *Discourses*, he explains how two balls suspended from two threads of equal length will keep the same beat, although one of the balls is in lead and the other in cork, and the first one will keep beating much longer than the second. He assures us that he has started the pendulums from a great variety of positions, up to almost horizontal ones, and that he has observed small and large amplitudes without detecting substantial changes, and he provides this as evidence that all bodies will fall at the same speed, whatever their weight, correction being made for air friction, which stops the light ball faster than the heavy one. For Galileo, the real impact of pendulum isochrony was to support his theory of motion, and this was much too important a conclusion to let mere facts get in the way; this probably explains his relative blindness to experimental results when they did not fit the theory. Galileo was neither the first nor the last to have let theory take precedence over experiments. The relation between theorists and experimenters in science has always been an uneasy one. Einstein himself, when presented with experimental results which seemed to contradict the theory of relativity, quipped famously, "The theory is good." Galileo also had a theory, and pendulum isochrony happened to be an essential part of it; even if the experimental results were not so hot, he had found a mathematical proof to reassure himself and convince skeptics.

Unfortunately, Galileo's proof is wrong. He first raises an interesting geometrical question. Suppose I want to build the fastest possible slide connecting a point A above the ground to a point B on the ground: what would its shape be? In other words, I let a weight slide from A without initial impulse all the way down to B, pulled by its own weight; clearly, if frictions are neglected, the time from A to B depends only on the shape of the slide, and I want the shape to be optimal in the sense that I want this time to be the smallest possible. I can guess what this optimal shape will look like: the slope should be steeper near A, so that the weight picks up speed at the beginning, and it should end up horizontally at *B*. Galileo went one step further, and asserted that the optimal shape is an arc of a circle joining *A* and *B*, which is wrong; from there he went on to prove that the pendulum is isochronous, which is wrong also.

As is often the case in mathematics, the mistake finally turned out to be fruitful, for it showed that the questions Galileo had tackled were more delicate than could have been assumed at first glance. Some of the greatest scientists of the time tried their hand at them. Does there exist a truly isochronous pendulum, the period of which would really be independent of the amplitude? What is the optimal shape for a slide between two given points *A* and *B*? These can be stated as two different problems in geometry. The first one consists of finding a curve such that the time it takes a point sliding on the curve (or a ball rolling on it) to reach the low point does not depend on the starting point. The second one consists in finding a curve between *A* and *B* such that a point sliding down from *A* will reach *B* in the smallest possible time. In fact, Galileo was right in thinking that both curves are the same, not circular arcs, as he believed, but arcs of a different curve, one of the most interesting discoveries of the seventeenth century, the *roulette*, as its inventors called it, or *cycloid*, as it is called nowadays. Pascal describes it like this: "The roulette is such a common curve that, after the straight line and the circle, none occurs as frequently; and it is so frequently under the eyes of everyone that one wonders why ancient writers have not considered it, since nothing about it can be found in their works: for it is nothing else but the path which a nail in a wheel travels in the air as the wheel rolls on the ground, taken from the moment the nail starts rising from the ground until the motion has brought it back down, the wheel having turned one full circle: assuming that the wheel is a perfect circle, the nail a point on its circumference, and the ground perfectly level."[2]

As Pascal points out, the ancient Greeks did not have the mathematical means of studying this curve; it cannot be constructed using only compass and ruler, and it cannot be described by an algebraic equation. To study the roulette, one had to wait for the new methods of calculus which were developed during the seventeenth century, and which culminated in the work of Gottfried Leibniz and Isaac Newton. The first important result is probably due to Roberval, who proved in 1638 that the area beneath one arc of the roulette is three times the area of the generating wheel. Galileo himself did not have the mathematical means to

2. *History of the Roulette* (1658).

1. THE ROULETTE This is a wooden roulette, built in the eighteenth century, and now in Florence, in the Museum of the History of Science. It serves to demonstrate the two main mathematical properties of the roulette. The first one is isochrony. A ball let loose on the railings will oscillate, by rolling down to the bottom, climbing up the opposite side, and falling back again to the bottom and up to its starting point. It behaves like a pendulum, but the period of its oscillations is independent of their amplitude. This is demonstrated by dropping simultaneously two balls on opposite sides from different heights and checking that they meet precisely at the bottom (the lowest point, in the middle). The second property is brachistochrony: rolling down the roulette is the fastest way to reach the bottom. This is demonstrated by dropping two balls simultaneously from the top right, one on the roulette and the other on the straight ledger: the one on the roulette reaches the bottom first.

There is a third mathematical property which can be used to construct roulettes. Mark a point M on a circle (for instance, put a spot of white paint on a bicycle tire), and roll the circle on level ground (ride the bicycle); the point M (the spot of paint) will delineate a roulette in the air. This roulette will be belly up instead of belly down (this is considered the right way to look at roulettes, so the wooden one in Florence is an inverted roulette); its top (highest point on the arch) is where the point M is on the top of the wheel, while its base is where the point M hits the ground.

identify the roulette as the true solution to the problems he had raised, although he was acquainted with it. That honor fell to Christiaan Huygens, and the Bernoulli brothers, Jacob and Johann, who showed that the inverted roulette (with the belly down) was the solution to both problems. Huygens showed in 1659 that the roulette-shaped pendulum would be truly isochronous (for a physical realization, let such a ball oscillate back and forth on an inverted roulette), and the Bernoullis showed in 1697 that a slide shaped like an inverted roulette would be the fastest route between two given points. Their proofs are landmarks in the history of the calculus of variations, a new mathematical discipline which we will describe at greater length in the forthcoming chapters, and which turned out to be an essential tool for the development of classical mechanics according to Galileo's ideas.

These theoretical advances would also lead to technological progress, with the result that, for the first time in history, scientists could look forward to new and accurate ways of measuring time. In 1637, in a letter to a Dutch correspondent, Galileo describes a clock, built on the pendulum principle, "so accurate that it is able to count time intervals, however small, without any error, in any place and any season." As usual, these

2. DESIGN FOR A PENDULUM CLOCK Drawing by Vincenzo Galilei and Vincenzo Viviani (1659), now at the Museum of the History of Science in Florence.

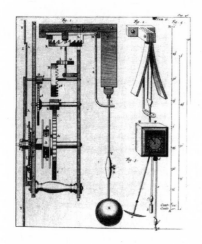 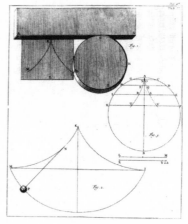

3. HUYGENS'S DESIGN FOR A CLOCK based on the isochrony of the roulette.
From his book *Horologium Oscillatorium* (Paris, 1673).

claims were somewhat premature; Galileo had no further interest in chronometry, and it is not known whether he ever had a pendulum clock built. There are some drawings for a pendulum clock by his son Vincenzo and his collaborator Vincenzo Viviani, but the instrument is quite rudimentary. It is Huygens who realized Galileo's dream and invented the modern mechanical clock.

Huygens devoted a large part of his life to the theoretical and practical problems connected with building pendulum clocks, and in 1673 he published a beautiful book on the subject called *Horologium oscillatorium*, that is, *On pendulum clocks*. With Huygens, theory and practice walked hand in hand; it was not enough for him to solve a problem mathematically, he also wanted to implement the solution with existing technology. For instance, as we have just seen, he had discovered that a roulette-shaped pendulum would be truly isochronous, that, is, that the period of its oscillations would not depend on their amplitude. A desirable property indeed, and one which the circular pendulum does not possess— but how could one build such a pendulum? A circular pendulum is simple enough: you just hang a weight from a string. But how does one impose a roulette-shaped trajectory to the weight? Huygens came up with a truly remarkable solution. He showed that, instead of letting the string dangle freely, it is enough to hang it between two curved blades, around which the string will partly wrap itself, thereby shortening its length. If the blades are appropriately shaped, the free end of the string will move along a roulette. Finding the appropriate shape for the blades is another mathematical problem which Huygens solved brilliantly, coming up, of all curves, with another roulette.

In 1657 Huygens was among the first to build a pendulum clock. In 1659 he built another clock, where the oscillator was no longer a pendulum, but a balance wheel, called back to its equilibrium position by a hairspring. But one has to keep the oscillations going; for that purpose Huygens invented a mechanical device called the escapement, which gives the balance wheel a small kick as it goes through equilibrium, but is not in contact with it at other times. The balance wheel and the escapement are standard features on all mechanical watches to this day. Despite his best efforts, however, which included investing a lot of his own money, Huygens could not realize his own ambition, namely to build a marine chronometer able to withstand the conditions at sea and keep time accurately enough to measure longitudes. Of course, there were major difficulties to overcome: the desired timepiece would have to be insensitive to changes in temperature (whereas a plain pendulum with a steel rod slows down by half a second per day when the temperature rises by one degree) and to changes in gravity across the Earth (whereas the same pendulum, brought from the pole to the equator, ceteris paribus, will slow down by 226 seconds per day). As we have seen, this objective would not be reached until half a century later, by John Harrison. But Huygens remains the founder of modern chronometry, one of the few examples in history of a major technological advance proceeding from scientists and theoreticians instead of practitioners and professionals.

To appreciate Huygens's contributions, it may be worthwhile to backtrack a few years, and to witness the efforts of Galileo's contemporaries to find a pendulum with a one-second beat. Indeed, when scientists around Europe decided to check Galileo's laws of motion for falling bodies, and to measure the associated physical constants, the need arose immediately for an accurate timepiece. The whole question boils down to finding the distance that a falling body drops during the first second after it is released. But how does one measure out one second, knowing that it is the 1/86400 part of a day? After Galileo, but before Huygens, the best answer consisted in building a pendulum which would beat exactly 86,400 times during one day. Mersenne got interested in the problem, and in 1636 determined that the correct pendulum would have a length of 3 feet (French style, with 32.87 cm to the foot). He then used his pendulum to measure the distance traveled during the first second of free fall, and found 12 feet, whereas Galileo had found a little more than half that amount, which goes to show how imprecise his experiments were.

Some years later, the Reverend Riccioli in Bologna tried his hand at the same problem; he built a pendulum of 3 feet, 4.2 inches (Bolognese style, with 29.57 centimeters to a foot) and, together with nine other Jesuit fathers, counted all the oscillations for a full day, namely May 12, 1642. They counted 86,999 oscillations and deduced that the correct pendulum would have a length of 3 feet, 3.27 inches. Riccioli then proceeded to find 15 feet for the distance traveled during the first second of free fall, more than twice as much as Galileo.

We are very far from these heroic times. The second today is defined, with an accuracy of one part in ten billion, as the duration of 9,192, 631,770 periods of the radiation emitted by an atom of cesium 133 during the transition between two hyperfine levels of the fundamental state, and there are timepieces which enable us to actually reach that level of accuracy. This may be Galileo's final vindication. The period of a real pendulum depends on its amplitude, although Galileo claimed the contrary; one full century of effort was needed before his ideas could be used to build a reliable timepiece; his own measurements were inaccurate; and his theories relied more on their inner coherence and his scientific prestige than on experimental evidence, if there was any. But today we have found an ideal pendulum, one which confirms Galileo's intuition, and we use it to measure time by counting the beats. This pendulum is not manmade; it is a wave of light, in accordance with the great intuition of Huygens, who was the first physicist to claim that light consists of waves and not of particles, and who made the first systematic study of waves and vibrations.

Even without its physical realizations, as chronometers or light waves, Galileo's ideal pendulum had considerable impact on the history of science. It turned time into a homogeneous quantity that could be divided into equal pieces, just like a length or a weight. These parts could be counted, and in this way various lengths of time could be compared and measured. From that moment on, mathematics entered the picture. Galileo discovered the mathematics of time, as the ancient Greeks discovered the mathematics of space.

The Birth of Modern Science

THE SCOPE OF MATHEMATICS extends far beyond geometry, and reaches to the very heart of reality. This was Galileo's great discovery, and he has recorded it for posterity: "Philosophy is written in that gigantic book which is perpetually open in front of our eyes (I allude to the universe), but no one can understand it who does not strive beforehand to learn the language and recognize the letters in which it is written. It is written in mathematical language, and its letters are triangles, circles, and other geometrical figures, and without these means it is humanly impossible to understand any of it; without them, all we can do is to wander aimlessly in an obscure labyrinth."[1]

Even today, four centuries after Galileo, one still wonders why mathematical concepts, related by equations and computations, are able to mimic and predict the behavior of physical systems in the real world. In 1960, the physicist Eugene Wigner wrote a famous paper, "On the Unreasonable Effectiveness of Mathematics in the Natural Sciences."[2] There should be a deep gap between the mathematical world, inhabited by ideas and concepts, with logical criterions for truth, and the physical world, made of objects and events, with no other criterion of truth than the evidence of our senses. How can these two worlds be connected? How can a mere calculation, or a logical argument, constrain the paths of galaxies or atoms? Inversely, how can consciousness and intelligence emerge in a purely material world?

And yet, it is so. Galileo's discovery marks the beginning of modern science. He has shown the way. All the scientists of the seventeenth century follow his path. Let us quote René Descartes, for instance, describing his

---

1. *Il saggiatore* (1623), chap. 6.
2. Wigner, "On the Unreasonable Effectiveness of Mathematics in the Natural Sciences," *Communications in Pure and Applied Mathematics* 13 (1960): 1–14.

years of study: "I was particularly fond of mathematics, where I found the arguments reliable and self-evident; but I did not notice at that point the true use of mathematics, for I thought they applied only to mechanical arts, and I wondered that nothing of greater import had been built on such firm and assured foundations."[3]

A few years later, Descartes himself would unify geometry and algebra, thereby creating modern mathematics, which turned out to be precisely the right tool to develop Galileo's ideas into a full-fledged theory of motion. Descartes' great invention, analytical geometry, translated every problem in geometry into a problem in algebra, which could be solved by computations. No longer was mathematics split into geometry and algebra, it became a unified theory. Mechanics is the study of motion, and at the end of the seventeenth century, as we shall see, it became analytical also, meaning that it was reduced to algebra: every problem in mechanics could be stated as a problem in algebra, the equations of motion could be written directly, and finding the motion meant solving the equation. In the three centuries since Descartes, geometry and mechanics were reduced to computation, and the proper procedure to solve a problem has been to try to solve the corresponding equation. Only with Henri Poincaré, at the eve of the twentieth century, did the limits of this approach become apparent.

The mathematics of Descartes were the proper tool to express and develop Galileo's ideas. But why are the right mathematics so powerful? An answer can be found in Bertolt Brecht's *Life of Galileo*,[4] in a lively and funny exchange between Galileo and Cardinal Barberini, soon to become pope under the name of Urban VIII:

BARBERINI. Are you so sure, my dear Galileo, that you astronomers are not just trying to make your own life easier? You have in mind circles, or ellipses, and uniform speeds, simple motions which your brains can understand. Imagine that it had pleased God to have His stars move like that. (*His finger draws in the air a very complicated path with an irregular speed.*) What would happen to your calculations?

GALILEO. Your Eminence, if God had constructed the world in this way, (*He retraces Barberini's path*) then he would also have constructed our brains that way (*He retraces the same path*), so that these paths would appear to us to be the simplest ones. I believe in the power of reason.

3. *Discours de la méthode* (1637), chap. 1.
4. Scene 10.

BARBERINI. I don't think the power of reason is good enough. See how he keeps silent. He is too polite to reply that he doesn't think the power of my reason is good enough.

Barberini, who by the way was an admirer and a friend of Galileo, raises an extremely interesting question: what do we mean by simplicity in science? What is a simple explanation, and why should physical laws be simple? In astronomy, for instance, there have been many different models of the universe, all claiming to be simple. During antiquity and the Middle Ages, going around a circle at constant speed was held to be the simplest of all possible motions. Moving along a straight line at constant speed was not considered possible, for the universe was thought to be surrounded by a huge solid sphere on which the stars were fixed, so that such a motion would necessarily end by hitting the boundary. It took a long time before a uniform linear motion was accepted as physically relevant, and even longer before it was considered to be simpler than a circular motion.[5] Even Galileo thought in terms of circular and uniform motions, and the first one to state explicitly that linear and uniform motions are the simplest of all was Descartes. He was also the first to state that a point traveling in space, free from any outside influence, would move at constant speed along a straight line.[6]

But for many centuries, circular uniform motions were considered the simplest of all, and for the sake of this false simplicity, astronomical systems, such as the one inherited from Ptolemy and transmitted by the Arabs, tried to combine such motions to approximate the actual motions of planets and stars in the sky: imagine a wheel carrying another wheel, itself carrying a third one which carries the planet, all rotating at different speeds. The end result is far from simple. After listening to a thorough exposition of the Ptolemaic system, King Alphonse X of Castile, nicknamed the Wise, is reported to have said that if Our Lord had done him the honor of consulting him before creating the world, he would have come up with some good advice.

For the sake of simplicity, Kepler did away with all these uniformly rotating wheels, and put the Sun squarely at the center of the universe. All the planets and the Earth itself rotate around the Sun with variable speeds, on elliptical trajectories: they accelerate as they get closer to the Sun, and they slow down as they pull away from it. For the sake of simplicity again,

5. From here on, "uniform" will mean "moving with constant speed"; "linear" will mean "moving along a straight line"; and "circular" will mean "moving along a circle."
6. This is usually referred to as "inertial motion."

Newton replaced Kepler's experimental laws by a deeper, mathematical one, for celestial bodies orbiting around the Sun, but for much more general situations as well. Unfortunately, this new-found simplicity turned out to be an illusion: Newton's law can lead to extraordinarily complicated motions. In the most important cases the equations of motion cannot be solved, and such basic questions as the long-term stability of the solar system remain unanswered to this day. Last but not least, it is also for the sake of simplicity that Einstein eventually did away with Galileo's idea of universal time. In Einstein's theory of general relativity, there is no longer any clear-cut distinction between space and time. There is instead a general geometry of space-time, giving rise to mathematical relations which include Newton's law of gravity as a special case. So we have gone from circular motions to elliptical ones, and then to mathematical relations of steadily increasing sophistication. Is that really the path to simplicity? Barberini's point is well taken: what do scientists mean by a simple explanation?

In the beginning of the fourteenth century, a Franciscan friar, William of Ockham, provided a first answer to this question. He stated a general principle which is known today as "Ockham's razor," because the implicit advice it carries is to cut out from our explanations everything we can remove without hurting the essence of the argument: "Concepts should not be multiplied beyond what is needed." Or again, "There is no point in doing with more what can be done with less." The argument by which the case at hand follows from general and accepted principles is always stronger than the one that devises new and untried explanations for a very particular situation.

The emperor Napoleon once asked the great astronomer Pierre Laplace, who had just published a vast treatise on celestial mechanics, what place he had left for God in cosmology. "Your Majesty," answered Laplace, "I found no need for this assumption." This is a beautiful instance of using Ockham's razor. Note that Laplace's answer is less obvious than it seems: Newton, for instance, needed this assumption, for he thought that the planets would eventually slow down on their path, or be diverted by small perturbations, so that once in a while God's hand would put them back on track. Not so Laplace: he found that all known motions of celestial bodies could be explained by applying Newton's law and performing the necessary calculations. To formulate this law one needed the Galilean framework of space and time, plus the mechanical concepts of mass and force, and Laplace very rightly felt that as long as they proved sufficient to explain natural phenomena, every other principle, such as the existence of God and his good will toward men, would be superfluous and hence metaphysical, beyond the domain of science.

Two centuries later, experimental evidence had accumulated that could not always be accounted for by Galilean mechanics; there was, for instance, a troubling discrepancy in the orbit of Mercury, the closest planet to the Sun. The planetary orbits change throughout the centuries, the main change being a very slow rotation around the Sun (such a rotation could not be seen if the orbits were circles, but, remember, they are ellipses). This rotation can be partly explained by the attraction of the other planets: the main force exerted on each planet is due to the Sun, but corrections must be made for other celestial bodies, which also exert an attraction on the planet. In the case of Mercury, however, this leaves forty-three seconds (of arc, not of time) per century unaccounted for, a trifling amount, but well within the accuracy of astronomical observations. New concepts were introduced by Einstein which accounted for this and other facts, and a new theory of space and time replaced the Galilean framework: this is the theory of relativity. So it finally turned out that new concepts were needed after Laplace, which probably would have surprised him very much, but Ockham's razor was still effective: Einstein introduced no more concepts than were absolutely necessary to account for known phenomena.

In fact, Newton himself says so in book 3 of his great work, published in 1687, the *Mathematical Principles of Natural Philosophy*. Here are the rules he lays down for the scientific method:

1. No other causes must be admitted but those which are necessary to explain the facts.
2. Thus, effects of the same kind must always, as far as possible, be attributed to the same cause.
3. Properties of objects which are not susceptible of variation, either of increasing or of diminishing, and which hold for all the objects one can experiment on, must be regarded as holding for all objects in general.
4. In experimental philosophy, the propositions which can be derived by induction from the facts must be regarded as true, or close to the truth, notwithstanding assumptions to the contrary, until some more facts confirm them entirely or show that there are exceptions. For no assumption can weaken an induction from experience.

Even today, after several centuries and scientific revolutions, there is nothing to change in these rules. Newton is a genius; he expresses deep thoughts in a concise way, very far, as we shall soon see, from the lyrical outpourings of Maupertuis. Newton knows that science is a chain connecting theory and experiments, and he pulls on the chain to check that it is unbroken. The four rules he states are a recipe for good science. The first two are an elaboration of Ockham's razor: do not introduce new concepts and principles if they are not needed, and use the old ones as far as they will go. The third rule is a principle of uniformity in nature, very similar to Ockham's razor: there is no reason to suspect that what you do not see is so very different from what you see. Even before one had seen the other side of the Moon, it would have been unreasonable to think it was covered with snow, or blue cheese. Newton gives other examples where this principle of homogeneity operates: "the breathing of humans and animals, a stone falling in Europe or in America, the light of a fire and of the Sun"; in each of these pairs, the phenomena must be thought of together. There is no reason to suspect that breathing is different in humans and in animals, or that stones will not fall in the same way in different places, or that light will behave differently if it is emitted by the Sun or by a fire.

Newton's last rule set the boundary between theory and experiment. A theory must arise by induction from the facts, it can never follow from other assumptions (for instance, that nature strives to achieve a certain purpose). It will be accepted only on provision, for the time being, until new experiments (not new assumptions) show its limits, in which case it will make way for a new theory. This—in 1687 yet—was an astonishingly modern view of science, very similar to the one Popper put forward in the middle of the twentieth century. Perhaps Newton put more emphasis on thrift, on using concepts and principles sparingly. If there was no regularity or uniformity in nature, if different explanations were to be found for the breathing of animals and for the breathing of humans, if one needed a theory for stones falling in Europe and another one for stones falling in America, causes would multiply without end, and no systematic knowledge would be possible.

It is a remarkable fact that this is not the case. Nature is no intellectual spendthrift, and a great many individual phenomena can be explained with very few general concepts and principles, connected by a set of logical relations and mathematical calculations. The best example, of course, is Newton's theory of gravity. Beside the Galilean framework of space and time, only three concepts are needed: acceleration, mass, and force. Once this is understood, the law of gravity is a very simple statement, namely that two bodies are attracted to each other, the force being proportional to their mass and inversely proportional to the square of their distance (if they are twice as far apart, the attraction is divided by four). This very simple statement is absolutely general: it holds throughout the universe, and applies to an extraordinary range of situations. With Newton's law, one can predict tides and lunar and solar eclipses, and position satellites over the Earth.

Of course, the mathematical arguments that lie behind all these predictions are far from easy. Newton's *Principia* is one of the greatest books of all time, and the reader cannot but be overwhelmed by the author's genius. Before Newton, Edmund Halley, Christopher Wren, and Robert Hooke all had formulated the inverse square law, mainly because of Kepler's observations connecting the distance of the planets from the Sun with the length of the year.[7] But only Newton had the mathematical talent to show that all the facts that Kepler had discovered about the planets followed as logical consequences from the inverse square law. Even today, with all the tools of modern mathematics at our disposal, it is no easy matter to perform the necessary calculations. Newton did it by elementary geometry, using only some properties of ellipses, which the ancient Greeks already knew.

How is it possible? How can physical reality be accounted for with a minimal set of rules, and seem compliant to logical arguments and mathematical calculations? For Newton, the answer is obvious: it is because God has created heaven and Earth. The whole planetary system has been constructed according to rules which are not beyond our intelligence, for God has also created us as a copies of himself, as it is said in Genesis; that is, as far as we are from God's perfection, we are still made after the same pattern. No wonder then that we understand this world and its rules, since we are of a kind with its maker: if we find that the planets' motions are explained by an inverse square law, this is simply because when God created matter, he decided that it would be subject to the inverse square law.

However, Newton is quick to point out that the inverse square law does not provide us with a full explanation: it tells us how motion proceeds; it does not tell us how it started. "The original, regular, position of these orbits cannot be attributed to these laws: the wondrous disposition of the Sun, the planets and the comets, can only be the work of an all-powerful and intelligent Being."[8] In other words, the world is a machine, and science gives us the blueprint. It will not tell us why the machine has been built in the first place, or what it is good for. Science is given a very modest role, simply to tell us how the machine works. Deeper explanations are to be sought elsewhere, if they are to be found at all. Newton himself spent a lot of time and

7. This is Kepler's third law: if the distance from the Sun is multiplied by k, the length of the year is multiplied by $k^{3/2}$.

8. *Scholium generale*, added to the second edition of the *Principia* (1713).

energy studying the prophecies in the Bible, and wrote over 300,000 words on the book of Revelation,[9] which probably was not the best use of his time. Let us enter the game by asking a question: what did God do after Creation?

Nicolas Malebranche, a disciple of Descartes, a strong believer by profession (he was a priest of the Oratoire order, and a very pious one), finds the answer in Genesis: "The seventh day, after having done all His work, He rested." After Creation, after having created heaven and Earth and started them moving, he rests, like a good engineer who takes pleasure in seeing the machine he has built function without a hitch. The mechanisms he has designed and put in place during the first six days keep the whole contraption going, and there is no longer any need for him to intervene. We may discover these mechanisms by lifting the hood and watching them in action, and we may reconstruct the blueprint of the machine. But, according to Malebranche, we will never be able to infer from the blueprint the goal of the engineer: only in holy scripture and the teachings of the church is God's purpose revealed. It is beyond the reach of mere science.

Newton and Malebranche are quite representative of their times. In the seventeenth and eighteenth centuries, the physical world was seen as a machine designed and set in motion by a creator. Having done his work, he stands by idly as the machine churns blindly along. This may seem, and is indeed, a very crude model of the world, especially if one remembers that there is a biological world along the physical one; animals, for instance, are considered by Descartes to be just another kind of machine. A senseless machine built by an intelligent being: this is a polarized world, and it will be very hard to pull the two pieces together. On the one side, an almost pure subjectivity, a mind that contains everything: the world is a temporary dream in God's eternal monologue. On the other side, a pure objectivity, a senseless machine, devoid of any consciousness, ready to sink back into oblivion as soon as it is no longer needed. Such pairs of alternatives abound in philosophy, they are all images of the original opposition between God and creation, and they are just as hard to overcome: soul and body, spirit and flesh, form and substance, *natura naturans* and *natura naturata*.

Another example, more important for our purposes, is the distinction between efficient causes and final causes, classic in ancient philosophy. Efficient causes give the blueprint of the machine, they describe the various mechanisms and their operations, while final causes start from the

9. See www.newtonproject.ic.ac.uk for an online edition.

engineer's purpose to explain the machine's design. This distinction is well explained by Plato, in his dialogue *Phaedo*. The question raised is whether Socrates, who has been condemned to death and is now in prison awaiting execution, should avail himself of an opportunity to escape. He teaches his disciples that there are two kinds of explanations for the fact that he is staying: one is that he has decided to on moral grounds (which is a final cause), and here is the other one (efficient cause): "Let us imagine a man who, while claiming that Socrates uses his intelligence to do whatever he decides to do, would proceed as follows when analyzing the causes of my actions. First, he would say, the reason I am sitting here right now is that my body is made of bones and muscles, that bones are rigid and connected to each other, while muscles can pull or let go, so that the muscles acting on the bones enable me to flex my arms and legs, until I find myself sitting here in front of you! Similarly, to describe our conversation, he would call upon other causes of the same kind, sounds being emitted, vibrations being carried to the ears, and many other such trifles, but he would not mention the true cause, which is that I have decided to sit here while the Athenians have decided to condemn me to death, so that each of us is doing what he thinks best."[10]

Indeed it is much more satisfactory to hear from Socrates why he is staying than to resort to an explanation in terms of muscles acting on bones. Unfortunately, we cannot question God in this way, and his purpose in creating the world (if there is a God and not several, and if he created the world) remains a mystery. So we have to do without: there can be no final causes in science, and we have to settle for the only remaining ones, the efficient causes. This means explaining everything in terms of nuts and bolts. Descartes, for instance, envisions a machine that would imitate the human body:

> I now want you to consider all the functions that this machine performs, such as the digestion of meats, the beating of the heart and arteries, the nurturing and growth of arms and legs, breathing, waking and sleeping; perceiving light, sounds, smells, tastes, heat, and other qualities, in the exterior organs of senses; the impression they make in the organs of common sense and imagination; the way they are retained or imprinted in memory; the interior motions of the spirit and the passions; finally the exterior motions of our body, which adjust so closely to the actions of objects which our senses perceive and to the passions and the impressions which they

10. *Phaedo*, 98.

encounter in our memory, that they imitate as perfectly as possible the behavior of a true human being. I want you to consider that all these functions follow in a straightforward way from the disposition of the organs in the machine, just as the motion of a clock, or of an automaton, follows from its weights and wheels.[11]

Animals are machines, and so is the human body.

This idea that the world is a machine, and that science should give us the blueprints, even if it cannot tell us the purpose of the machine, will lasted until modern times. There is nothing surprising about this: during the Renaissance, science was closely connected with technology (which was not the case in classical antiquity), and the industrial revolution made the ties even closer. The greatest scientists of the times, Galileo, Huygens, Pascal, not to speak of Leonardo da Vinci, were all engineers, and they built instruments as an integral part of their scientific work. For an engineer, to understand something is to be able to build it; whatever you have built yourself you understand perfectly. If the world can be understood, it is because it is a machine, built by the greatest of all engineers. The world is a clock, and by seeing the clock you know there is a watchmaker. God's work can be seen, and his character inferred, from the wondrous arrangement of the physical world rather than in the blood-thirsty histories of the Bible: he is, above all, a rational being. These ideas reached their culmination during the French Revolution; temples were built to the goddess Reason throughout the nation, and on the façade an inscription read "The French people recognizes the Supreme Being and the immortality of the soul."

Galileo raised reason to a pinnacle. All the scientists and philosophers of the time agreed that reason stands above everything: it is not created, as the world is; on the contrary, it presides over creation. Even for true Christians, like Malebranche, "sovereign reason is coeternal and consubstantial with God." In other words, God is not rational, or reasonable: he is reason itself. Mathematical and logical truths are not created: they are part of divine nature, and God could no more change them than he could change himself. Brecht puts in the mouth of Galileo the idea that God could have created quite a different world, inhabited by different humans who would not consider ellipses and circles to be the simplest possible paths. This idea is truly a modern one, quite foreign to the seventeenth and eighteenth centuries. For those times, God does not

11. Annie Bitbol-Hesperies and Jean-Pierre Verdet, eds., *Le monde, l'homme* (Paris: Editions du Seuil, 1996), conclusion.

make the rules of logic or mathematics, they are part of himself, and are the same in every possible world.

We now run into a very interesting question, one that came into fashion when Galileo discovered that Jupiter had satellites, and that there were mountains on the moon: could there be other worlds? Is this the only one possible, and if not, do others exist? Indeed, if God created one world, he presumably could create more, perhaps with different rules. He is bound to keep the same rules for logic and mathematics, but what about physics? Could God create a world where the inverse square law was replaced by another one, an inverse cube law, for instance? In this hypothetical world, the attraction between two bodies would be inversely proportional to the cube of their distance, so that if they are twice as far apart, the force would be divided by eight (instead of four). Such a world would be as logical and mathematical as ours; has it been created, and if so, where? If not, why?

One would, of course, answer that such investigations are straying well away from the path of science. But no, not in those times, where new worlds were continuously discovered, not only by exploring the Earth, but by raising one's eyes. Galileo was the first to direct one of the newly discovered field glasses to the nightly sky, and he found scores of new heavenly bodies, some quite close to us, like the satellites of Jupiter, some very far, like stars of low magnitude. Even very familiar objects had revealed totally unsuspected and remarkably original features: the Moon had mountains and seas, the Sun had spots, Saturn had rings. These were no mere lamps moving around the sky; these were entire worlds, very different from the Earth, and sometimes larger. As astronomers were exploring the heavens, others were looking in the other direction, and using microscopes to investigate the smaller parts of our world. They saw insects and worms turn into gigantic monsters, and probing even deeper they found entire populations of minuscule creatures living happily far below the reach of our sight. The world is full, at every scale, and every scale ignores the higher and lower ones.

The philosophical consequences of these discoveries are immense: humans were born in a finite cosmos, surrounded by a sphere carrying the stars. At the time of the Renaissance, they emerged into an infinite universe. Perhaps they were not alone, for there was no reason to believe that all the new worlds astronomers were discovering were so many deserts. If they turned out to be inhabited, another Copernican revolution would be needed: humankind would no longer be at the center of creation, just like the Earth no longer was at the center of the universe. It could also be the case that the worlds discovered beneath us, in the smaller dimensions, had some similar surprise in store: since there was an abundance of life at all scales, why should intelligence be limited to our size?

There is a short story by Voltaire that captures the philosophical mood of the times. On the star Sirius lives a benevolent and educated giant, named Micromegas (small/large, in Greek), who wants to see the world. He finds a travel companion, much smaller than himself (although still extremely large by our standards), but overcomes his prejudice with the admirable thought that "a thinking being should not be dismissed for being only six thousand feet tall." They land on the planet Earth, and at first they find no life on it, for the scale of men and animals is far below what they can see. Fortunately, they have the bright idea to observe Earth through a microscope, and then of course they start seeing small creatures running and swimming around. In fact, they hit upon the ship carrying Maupertuis back from Scandinavia, and of course Voltaire does not miss the opportunity to ridicule his old enemy one more time.

There were many more stories, novels, and treatises dealing with other possible worlds. In Great Britain, John Wilkins' book *Discovery of a New World, or a Discourse Tending to Prove That It Is Probable That There May Be Another Habitable World in the Moon*, published in 1638, met with a huge success. In France, Fontenelle wrote *Discussions on the Plurality of Worlds*; when he died in 1757, the book had been reprinted thirty times. We should also quote Cyrano de Bergerac, who is famous for other reasons, and Pierre Borel, who wrote in his *New Discourse Proving That Celestial Bodies Are Inhabited Earths*, published in 1657, that "humans should stop behaving like ignorant peasants, who have never set foot outside their small village, and yet believe firmly that nothing in the world could be as magnificent." Good advice, and still valid today.

The major work in all this literature probably is the book by Tommaso Campanella, *Apology of Galileo*, written in prison and published in 1616. He puts all the astronomical discoveries of Galileo in perspective, and shows how they fit the picture of a single, well-organized universe. For instance, it is hard to imagine that Earth is floating around in empty space, with the Moon circling around it, but is much easier when we see Jupiter doing precisely that, with not one but five satellites. Similarly, the phases of the Moon are no longer an isolated phenomenon: Galileo has discovered that the planet Venus has phases as well, so that we can identify the common cause, namely the Sun. Mountains are known on Earth, but there are also mountains on the Moon, and we can recognize them by the shadow they project. We see the same phenomena in different surroundings, so that our experience on Earth can be extrapolated to all the new worlds that have been discovered. A few years later, Isaac Newton would vindicate Campanella's vision by showing that a universal law, valid throughout the universe, can account for all these observations. We do not have different laws of nature for different planets or different scales: they are the same throughout. So, in fact, they are all part of the same world, and we are back to the same question: are there any other worlds? If not, why is this the only one to exist?

Gottfried Wilhelm Leibniz (1646–1716) may be the only philosopher who ever tried to answer this question in a precise way. He is certainly one of the greatest intellectual figures of the age, on a par with such giants as Newton and Spinoza. As a scientist, he is remembered as the inventor of differential calculus, in spite of a quarrel with Newton about priority, which he largely seems to have won. Modern mathematicians still use the concepts he introduced and the notations he chose. His work in philosophy is now seen through the distorting prism of quarrels which happened long after his death, and he is chiefly remembered as the proponent of the much ridiculed idea that we are living in the best of all possible worlds. A bold statement indeed, but what Leibniz actually means is far from naive, and deserves some attention.

Leibniz begins by defining precisely what is meant by a "possible" world. Let us start from the fundamental duality we described earlier, the face-to-face between God and creation, or subject and object. There are many ways in which God can create a world, but he has to abide by some logical principles, noncontradiction for instance: "to be" means to be something, and if you are something, you cannot simultaneously be something else. Even God cannot create something which is at once a triangle and a circle. A triangle is a triangle: it consists of three vertices joined by three straight sides. A circle is a circle: it consists of all points at the same distance from the center. There is no way a triangle could ever be a circle, or a circle a triangle. Such clear-cut divisions might not hold if one wandered outside the realm of mathematics; it is not so obvious, for instance, that one cannot be simultaneously good and evil, or adult and child. But Leibniz thinks as a Cartesian; that is, he draws from well-defined concepts, free from any ambiguity. The noncontradiction principle is then reduced to a tautology, a logical statement that is true for purely formal reasons, independent of any empirical content. Whatever a statement "S" has to say, "S" and its opposite cannot be true together: such is the strength of logic.

According to Leibniz, this strength binds even God. Everything in existence must satisfy the noncontradiction principle, and the identity

4. GOTTFRIED WILHELM LEIBNIZ
(1646–1716).

principle as well: whatever S is, it must be itself. S is S. In Leibniz's philosophy, existence is just a way of checking out the noncontradiction
principle and the identity principle. Any concept that satisfies both is
"possible," and God can bring it to existence. But not every possible
concept exists: it is up to God to decide which ones he will endow with
existence. In so doing, he faces a superhuman problem, which is to
choose them in a consistent way: inserting a new series of events into
the world requires not only that it entail no contradiction within itself,
but also that it fit into the already constructed chain of being, so that it
entails no contradiction with existing events. In Leibniz's terminology,
the world consists of things which are not only possible by themselves,
but also "compossible" with the others. However, between possibilities
which come to existence and possibilities which don't, there is no other
difference than this divine election, which is an outside event and does
not affect their true nature. This may be compared to God sending some
of us to heaven and others to hell: such outcomes can be understood as
logical consequences of our personalities, but they are not part of them.
All possible things coexist in God's mind, which, in Leibniz's words, is
"the land of possible realities."[12]

12. Letter to Arnauld, July 14, 1686.

Each of these possible realities is entirely encompassed by the principles of identity and noncontradiction. For instance, there was always present in God's mind an Ivar Ekeland typing these lines in Chicago on a fine summer night in 1998. This particular idea of Ivar Ekeland also encompasses many other things: childhood sicknesses, scientific papers, and cruises in the Aegean. In fact, it is nothing else than the full story of Ivar Ekeland, from conception to death, down to the smallest detail. The life I live is nothing but the gradual unfolding of that particular idea, which, however, is as immediately accessible to God's mind as the idea of a circle is to mine. There are also all the possible Ivar Ekelands which never existed—the one who was born a day later, the one who was run over by a car when crossing the street on his way to school, the one who decided to live in Norway. They are all there, infinitely many of them; each of them would have been called Ivar Ekeland, but would have done something different with his life. They are all present together in God's mind, but he bestowed existence to one only.

This way of looking at existence provides an elegant answer to the age-worn question of predestination: how can God create me, and create me free? If I end up in hell, should I not blame him for not having created me a better man? Suppose I murder someone; since nothing is hidden from God, he certainly knew in advance I was going to commit this crime. But if so, I was no longer free not to commit it: I may have imagined, at the time of the deed, that it was up to me to fire the gun or not, but I was mistaken; it had all been decided long ago. From the very beginning, it was preordained that I would kill that person. I am like Oedipus, who was doomed from birth to kill his father and marry his mother. I am luckier than he was, for the crime that was imposed upon me is not as heinous, and less so, because I was never warned, whereas an oracle told him his fate, and he could take steps to avoid it. If I had been truly free, the choice would have remained open till I actually pulled the trigger. The fact that the result was known in advance proves that the choice was not mine, and I should not bear responsibility for it.

This is basically the example that Leibniz gives in his *Theodicy*. He connects it with an episode in Roman history, the rape of Lucretia, which was so well known at the time that Shakespeare had made it the subject of a lengthy verse narrative. The villain in this particular story is one Sextus Tarquin, and his deed resulted in the death of Lucretia, who killed herself in shame. In vengeance, her family then succeeded in overthrowing the reigning king, Sextus's father, and that was the end of the monarchy in Rome. So the rape of Lucretia, an evil act in itself, changed history in a positive way, since it replaced a tyrannical and corrupt regime with a better one, the Roman Republic, which went on to conquer the whole world. This is an example of the mixed consequences that a single action may have, and it may be argued that in that particular case the good overrode the bad. But Leibniz's concern is with Sextus himself. He has him present the argument that, since God knew in advance that the Roman Republic had to come into existence, he, Sextus, had really no choice but to rape Lucretia, and should not be held accountable for that crime. Leibniz then points out that, at that very moment, there are many other Sextuses present in God's mind, most of whom would have behaved quite decently at that juncture, out of their own free will. The particular Sextus who is complaining exercised free will, just like the others: his misfortune is that he was the one that God chose to bring into the world. He decided to rape Lucretia, and God decided to give him existence.

So every possible reality, once God gives it existence, will reveal its own identity. Its life will unfold it slowly, whereas God encompasses it at a single glance. God's mind is like the library which Jorge Luis Borges describes in a famous tale,[13] with infinitely many books shelved in infinitely many stacks. Each of these books contains a full biography of a possible person, Ivar Ekeland, say, but they are no help to me as I browse around, because I don't know which is—or will be—the true one. I can open one of the books and check whether the events described until today are correct, but as for the events which are recorded but are yet to occur, I have no clue. There are many other libraries, each of them containing possible biographies of other individuals. God has read them all, and he will bring to existence one book from each library, and all these biographies have to fit with each other.

This is quite an undertaking. As I said before, it is not enough that each of these biographies is internally coherent, it must also agree with all the other ones which have been chosen when they describe common events. If two biographees meet, both accounts of their encounter must agree to the last detail. There are also other realities to worry about— plants, animals and things—all of which have to be chosen among infinitely many possibilities, and all of which have to fit with all the other ones. The whole world has to be created in a coherent way, and, as everyone knows, the devil is in the details. One finds some attempts of this kind in works of fiction: Frank Herbert, for instance, or J. R. R. Tolkien,

13. "The Babel Library," from *Fictions*, 1944.

have succeeded in creating imaginary worlds that stick in our minds because of the painstaking care with which every detail is woven into the whole. Tolkien conjured up the whole history and folklore of Middle-earth in support of the adventures of Bilbo the Hobbit, and Frank Herbert went at great length to imagine what kind of culture and technology would develop on a waterless planet like Dune.

If we could push Tolkien's or Herbert's idea to the limit, describing not only Hobbits or humans but all the other forms of life, as well as the natural laws which would be valid in these particular worlds, we would get instances of what Leibniz calls a "possible" world: a complete and coherent set of possible realities. There may be many such possible worlds, even outside the realm of fiction. One could imagine, for instance, that the Pearl Harbor attack had never occurred, and that the United States did not enter the war in 1941. One could also imagine that evolution had taken another route, sparing the dinosaurs, which would still be around today as the dominant species on Earth. Finally, one can wonder what kind of life may have developed on planets lost in interstellar space, far from any sun, but endowed with water and volcanoes, and hence oceans and continents.[14]

Why did God choose this particular world to exist? Many who experience it find much to object in the way it is run. Let us look up Leibniz's answer in the *Monadology*. It is a concise exposition of his theory, in ninety propositions, and it is the only philosophical book he published in his lifetime.

> 53. Now, since in the divine ideas there is an infinity of possible universes of which only one can exist, the choice made by God must have a sufficient reason which determines him to the one rather than to another.
> 54. This reason can be found only in *fitness*, that is, in the degree of perfection contained in these worlds. For each possible has a right to claim existence in proportion to the perfection it involves.
> 55. This is the cause for the existence of the best, which is disclosed to God by his wisdom, determines his choice by his goodness, and is produced by his power.[15]

So this is it: the present world has been chosen to exist because it is the best of all possible worlds. Tarry a moment: how "the best"? From what proposition 54 says, it has to be the most perfect. And

14. "Life-sustaining Planets in Interstellar Space?" *Nature* 400 (1999): 32.
15. Translations by Paul Schrecker and Anne Martin Schrecker, in *Monadology and Other Philosophical Essays* (Indianapolis: Bobbs-Merrill, 1965).

what is perfection? That's what the next three propositions set about to explain:

56. This *connection* of all created things with every single one of them and their adaptation to every single one, as well as the connection and adaptation of every single thing to all others, has the result that every single substance stands in relations which express all the others. Whence every single substance is a perpetual living mirror of the universe.

57. Just as the same city regarded from different sides offers quite different aspects, and thus appears multiplied by the perspective, so it also happens that the infinite multitude of simple substances creates the appearance of as many different universes. Yet they are but perspectives of a single universe, varied according to the points of view, which differ in each monad.

58. This is the means of obtaining the greatest variety, together with the greatest possible order; in other words, it is the means of obtaining as much perfection as possible.

So perfection consists of two things: variety on the one hand, that is the inexhaustible profusion of natural phenomena; and order on the other, that is the interrelatedness of all things and the basic simplicity of natural laws. Leibniz sees variety and order as two sides of the same coin; in a 1679 letter to Malebranche, he explains: "We also have to say that God makes as many things as it is possible, and the reason why he has to look for simple laws is precisely that he must accommodate as many things as it is possible to fit together: if he were using other laws, it would be like building a house with round stones, which would lose as much space as they fill." However well this thought is put, it is not clear how one can achieve simultaneously the greatest possible variety and the greatest possible order: one would rather expect that some kind of compromise would have to be reached between these two criteria. Leibniz does not follow up on this point, and does not seek to formulate some kind of quantitative criterion to be maximized. He is more inclined to qualitative arguments, which seek an overriding harmony in the abundance and variety of the universe.

To understand Leibniz's world, think of constituting an orchestra. First, there must be instruments, and musicians able to play them; these are all the possible realities, waiting to be called forth. But a choice must be made: not all can play. There are certain rules to follow as to the instruments. To get a balanced sound, and the musicians must learn to play together; this is where one checks that all the individual choices that were made fit into a coherent ensemble. Every such orchestra is a possible world, and now comes the tough problem: choose the best one, for it is the one that will be created. It is doubtful that quantitative criteria will be devised to rank orchestras, or even that there will ever be general agreement as to which orchestra is the best at any given time.

Certainly Leibniz does not think of anything so crude when he talks of this world as being the best of all possible ones. He does not have a mathematical formula in mind. Neither does he put human happiness in the forefront: there is scarcely any mention of it. Happiness may play a role, insofar as it is a component of the universal harmony, but this role is not essential, or even important. Of course, it all depends on what constitutes happiness; Leibniz belongs to this category of philosophers who claim that happiness lies in contemplating the wonders of God in his creation, an idea that is certainly far away from the everyday concerns of most human beings. All in all, to say that this world is the best of all possible worlds does not necessarily imply that it is a pleasant one to live in. In fact, since it has to be both as varied and as orderly as possible, it must accommodate extremely different beings, living together under simple laws; this can only lead to compromises, which cannot all be expected to be advantageous for all concerned.

We are far from the crude philosophy "All's well that ends well" incorrectly attributed to Leibniz. We are also far from the mechanistic conception of the universe which was shared by Descartes and Newton. Leibniz does not see the world as a machine; he sees it, in his own words, "as a garden full of plants or as a pond full of fish." He is a naturalist, whereas the others are engineers. He is not like Galileo, who raised his telescope to the stars and explored the infinite universe above him; he is like Antony van Leeuwenhoek, the inventor of the microscope, who looked at the infinite universe below him and explored the worlds contained in a single drop of water. Indeed, let us put side by side a quotation of a letter from Leeuwenhoek to Robert Hooke (1676):

> It is as if one saw, with the naked eye, small eels writhing against each other, and all the water was alive with these minuscule animals; and of all the wonders I have observed in nature, this is the most wonderful of all.

and three more propositions of the *Monadology*:

> 67. Thus every portion of matter can be conceived as a garden full of plants or as a pond full of fish. But every branch of the plant, every limb of the animal, every drop of its humors, is again such a garden or such a pond.
> 68. And though the soil and the air in the intervals between the plants of the garden is not a plant, nor the water between the fishes a fish, yet these

intervals contain again plants or fishes. But these living beings most frequently are so minute that they remain imperceptible to us.

69. Thus there is nothing uncultured, sterile or dead in the universe, no chaos, no disorder, though this may be what appears. It would be about the same with a pond seen from a distance: you would perceive a confused movement, a squirming of fishes, if I may so, without discerning the single fish.

This is Leibniz's philosophical inheritance. Fifty years after his death, largely by Voltaire's doing, it will be mixed up with a very different philosophy, Maupertuis' quantitative and mechanistic metaphysics, according to which, "in ordinary life, the sum of benefits exceeds the sum of misfortunes." How such a philosophy came to be, and how it was confused with that of Leibniz, will be the subject of our next chapter.

The Least Action Principle

IN JUNE 1633, Galileo was condemned. After a six-month trial, the Inquisition tribunal declared him "vehemently suspect of heresy, for having held forth and believed a false doctrine, contrary to the Holy Scriptures; to wit, that the Sun is the center of the world and does not move from East to West, that the Earth moves and is not the center of the world, and that one may hold probable and defend an opinion after it has been examined and declared contrary to the Holy Scriptures." In the course of the trial, it had been proved beyond doubt that the accused had not been content with supporting these abominable opinions; he had also done everything in his power to propagate them as widely as possible, for instance by using the simple trick of writing in the vernacular. As the sentence puts it, "Not only does he provide the Copernican opinion with new weapons, which no foreigner has ever thought of, but he does so in Italian, the language which is the most likely to bring to his side the ignorant people, those among whom errors find the most fertile ground."

Wouldn't it have been wiser and fairer, indeed, for a Christian gentleman, to put forth his arguments in Latin, thereby restricting access to prudent and educated people, well acquainted with the holy scriptures, and the Church Fathers, better equipped to find the dangers lurking in new ideas, and less likely to be infected by them? The sentence goes on to reject Galileo's defense, according to which he was merely setting forth a mathematical theory without practical consequences: "The author claims to have discussed a mathematical hypothesis, but he attributes to it a physical reality, something that mathematicians would never do."

It was not a warning; it was a condemnation. On June 22, 1633, Galileo, in a penitent's gown, knelt before the cardinals of the Sacred Congregation of the Holy Inquisition, and pronounced a public retraction: "With sincere heart and unfeigned faith, I abjure, curse and detest the above errors and heresies, and I swear that in the future I will never again say or assert, orally or in writing, things which may arouse similar suspicions, but if I ever encounter some heretic or anyone suspect of heresy, I will turn him over to this Holy Office." He was condemned to ending his life in detention, first in Sienna, and later at his villa in Arcetri, near Florence, where he remained under house arrest until his death, a blind and broken man, in 1642.

The blow was felt throughout Europe, where a thriving scientific community had developed. Long before the Internet, information flowed through daily exchanges of letters; people like Mersenne maintained a vast correspondence throughout Europe, and served as information hubs, disseminating news, recording progress, and distributing problems to be solved. Galileo was a towering figure in this world, everywhere his discoveries were known, his books quoted. He was the first to have turned a telescope toward the night sky. He discovered that Jupiter has satellites like the Earth, that Venus has phases like the Moon, and that the Moon has mountains and seas like the Earth. He also discovered that the shape of Saturn changes, from a circle to an oval; his instruments were not sharp enough to separate the rings from the body of the planet. The books inherited from antiquity make no mention of these facts, but anyone could ascertain them by pointing a telescope to the sky, without first having to learn Latin or Greek. It was a triumph of the experimental method over mere bookish knowledge; from then on, research would interrogate nature rather than tradition, and the idea of scientific progress, constantly extending the boundaries of knowledge, took hold. One century before, Martin Luther had freed believers from the shackles of tradition and had empowered them with the right to read and understand scripture by themselves; Galileo taught individuals to see through their own eyes, and to seek the truth in nature rather than in the writings of ancient philosophers. In addition, Galileo was an influential person in the political world. He was a personal friend of pope Urban VIII, whom he knew as cardinal Maffeo Barberini, and who expressed his admiration for Galileo's scientific work in several letters written over the years. Galileo had been honored by the Republic of Venice, by the Duke of Florence, to whom he dedicated the newly discovered satellites of Jupiter, henceforth known as the Medicean stars. He had powerful friends in the Roman Curia who had fended off several attempts against him. In 1616, for instance, even when Copernicus's opinion was condemned and his book prohibited to the faithful, Galileo got off with a warning "to completely abandon that opinion, and in no way to hold it, or to defend it, or to teach it."

But this time, things proceeded quickly. Galileo's *Dialogue on the Two Greatest Systems of the World, the Ptolemaic and the Copernican*, appeared in 1632; the inquisitor in Florence ordered a halt to its diffusion, and in October Galileo was summoned to Rome. He made the trip in January 1633, appeared before the court on April 12, and two months later, on June 22, was sentenced. This was a lesson for others as well. In November of the same year, Descartes learned of Galileo's condemnation, and immediately decided not to publish his magnum opus, the *Treatise of the World, or Of Light*. He had been working on it ever since he had settled in the Netherlands five years earlier, and it lay ready to be sent to the printer. This was a momentous decision, for the *Treatise* was to have been the linchpin of his philosophy, the center from which Descartes' achievements in science and metaphysics would appear as a well-ordered whole. In addition, the experience exaggerated Descartes' natural tendency toward prudence, and toward protecting his ideas by ambiguous formulations. From then on, according to his own account, he would proceed "under a mask." Gone was the mystical enthusiasm of that night in Ulm where "the foundations of a marvelous science" were revealed to him.[1] The epitaph on his grave still carries an echo of that initial impulse: "Taking advantage of his Northern sojourn[2] to connect the laws of mathematics with the mysteries of nature, he was bold enough to hope that one could open both secrets with the same key."

In 1637, Descartes decided to give it another try. He published anonymously a book containing three short scientific treatises, *Optics*, *Meteors* (meaning large-scale natural phenomena, such as the rainbow), and *Geometry*. There was also a general introduction which, isolated from the rest, was to become one of the most important texts in the history of philosophy: it is the celebrated *Discourse on Method*, in which Descartes lays the ground for his philosophy, and described the method by which one might "find the truth in the sciences." Following the example set by Galileo, the book was written in French rather than Latin. Strangely enough, the *Discourse* does not seem to have attracted much attention at the time, but the scientific treatises did. Much of Descartes' correspondence at the time was devoted to the discussions and controversies surrounding them.

1. November 10, 1619; "Ut comoedi, moniti ne in fronte appareat pudor, personam induunt, sic ego hoc mundi teatrum conscensurus, in quo hactenus spectator exstiti, larvatus prodeo." *Cogitationes Privatae* (1619), in *Oeuvres de Descartes*, ed. C. Adam and P. Tannery (Paris, 1897–1913), 10:213.4–6.
2. Descartes spent most of his active years in the Netherlands and died in Sweden.

From that time on, these four fragments of the *Treatise of the World* have led separate lives. The *Discourse* is no longer seen as an introduction to a scientific treatise, but as a self-standing work in philosophy. The *Geometry* created a new science, mathematics, by unifying arithmetic and geometry. Up to Descartes, these were distinct endeavors: arithmetic dealt with numbers, integers, or fractions; geometry with shapes, in the plane, like squares and circles, or in space, like cubes or spheres. Descartes hit upon the idea of representing each point in the plane by two numbers—the so-called Cartesian coordinates—and each point in space by three. By this method, every problem about shapes could be translated into a problem about numbers and vice-versa, so that geometry and arithmetic were seen as two sides of the same coin: the *Geometry* was the first treatise in modern mathematics. Similarly, the *Optics* and *Meteors* constituted a full-blown theory of light, where Descartes proves the law of refraction from first principles and uses it to explain rainbows. These are works in physics, to be read and understood on their own, while the overall connection with mathematics and philosophy has been forgotten.

This would probably not have happened had Descartes been able to publish the *Treatise of the World* in due course. An essential part of his message has thus been lost, for unity is central to his way of thinking. In a set of notes which were obviously meant for personal use, and which were found in his papers after his death, he writes, "All sciences are nothing but human wisdom, which remains one and the same, however different the objects to which it applies itself may be, and which is no more changed by these objects than the light of the sun by the variety of things it illuminates."[3] Of course, this human wisdom is but the reflection of God's own wisdom, and the ability to grasp the rules by which he has created the world, chief among which are the rules of mathematics. In an unpublished section of his *Treatise of the World*, Descartes writes, "I will be content with warning you that, besides the three laws I have explained, I will assume none other than those which follow unfailingly from these eternal truths on which mathematicians are accustomed to rest their most certain and most transparent proofs; these truths, I say, according to which God himself has taught us that he had arranged all things in number, weight and measure, and the knowledge of which is so familiar to our souls that we cannot but deem them infallible whenever we perceive them distinctly, nor doubt that, if God had created several worlds, each of them would be in every respect

3. *Rules for Directing the Mind*, published 1701 but probably written around 1628; rule 1.

as truthful as this one. In this way, whoever is sufficiently alert in examining the consequences of these truths and of our rules will be able to recognize the effects from their causes, and, if I may express myself in scholarly terms, will have *a priori* proofs of everything that could be produced in this new world." In other words, the laws of nature are nothing but the rules by which God has built the world; we have access to them because he is truthful, and he is bound by them to such an extent that, if he has created other worlds, they would still apply and we could reconstruct these worlds from basic principles, with minimal recourse to experimentation.

A modern scientist, wondering about "the unreasonable effectiveness of mathematics in the physical sciences," may sympathize with these ideas. Nowadays, however, we are much more aware of the fact that the best proof in the world is worth no more than its premises: every scientific theory is transitory and provisional, in wait for a better one, and accepted only as long as the experimental results conform to its predictions. Descartes, on the other hand, believed that science rests on eternal truths. As a consequence, he held experimental results in low esteem, thinking them to be error-prone (not a wild claim in his time) and less trustworthy than sound argumentation. His was a normative science, telling nature what it was supposed to do, not a positive one, investigating what it was actually doing.

After Descartes' death in 1650 his ideas were carried on by his disciples, most notably Claude Clerselier (1614–1684), who was responsible for finally publishing the *Treatise of the World* in 1677. Throughout the seventeenth century, and well into the eighteenth, the Cartesians fought for the master's ideas, and against the onset of Newtonian physics. The quarrel with Fermat in 1662–1665, and Maupertuis' expedition to the North in 1736–1737, which we shall describe presently, can be seen in perspective as the opening and the closing battles in this long, drawn-out struggle, ending with the final triumph of Newtonian physics. This shift toward a more experimental science is well exemplified by the personal evolution of another great physicist, Huygens: "When I read [Descartes'] *Principles*[4] for the first time, I was under the impression that everything was proceeding smoothly, and I believed, when I encountered some difficulty, that it was my fault, not to be able to grasp his thought fully. I was then but fifteen or sixteen years old. But having since discovered in that book things which are patently wrong, and others which are extremely unlikely, I have much reconsidered my previous position, and today, in all

4. *Principles of Philosophy* (1644).

of his physics, or metaphysics, or meteors, I find almost nothing to which I can subscribe as being correct."[5]

Let us now open Descartes' treatise *Optics*, part of the book published in 1637. In the first chapter, we are told that, in a homogeneous, transparent medium, light propagates in a straight line, called a ray. Without saying in so many words that light consists of particles, Descartes compares it to a tennis ball which rebounds at a different angle if it is sliced instead of lifted.[6] The second chapter deals with refraction. This is the phenomenon by which the rays change direction when light goes from air into water, which is the cause of many optical illusions. A stick, one end of which is dipped in a pond, appears to be broken at the surface of the water. Do not try to harpoon a fish from the bank of the river: it is not where it seems to be.

Descartes compares light traveling from air to water to a tennis ball which is accelerated in the vertical direction as it crosses the surface, the horizontal speed being unaffected. From this he derives the celebrated "sine law," $\sin i = n \sin r$, where i is the angle of the incoming ray with the vertical (in the air), r is the angle of the outgoing ray with the vertical (in the water), and n is simply the acceleration factor, that is, how much faster light travels in water than in air:

$$n = \frac{\text{speed of light in water}}{\text{speed of light in air}}$$

This number n, called the *refraction index*, turns out to be about 1.33.

This law had already been discovered by a Dutchman, Willebrord Snell, in 1620, but the interesting point here is that Descartes actually proves it. His argument, as one goes through it, is mathematically and logically correct, so the conclusion must follow, provided however that one grants the premise, namely that light is accelerated as it enters the water. We are then faced with a factual question: is it true that light travels faster in water than in air? It would be wrong to believe that this is the case, just because the conclusion Descartes draws from it, namely Snell's law of refraction, is correct: many a correct statement has been drawn from false premises. The question was not settled until 1850, more than two centuries later, when Léon Foucault and Hippolyte Fizeau measured the speed of light in water. In the meantime, it was a matter of opinion,

5. Letter to Pierre Bayle, February 26, 1693.
6. Tennis did not exist in the modern version, but its ancestor, the French *longue paume*, was a popular racquet game in Descartes' times, and is still played today.

the majority (including Isaac Newton) holding the opinion that light, like sound, travels faster in water than in air. Very few people did not, and the first among them was probably Pierre de Fermat (1601–1665), another of the great mathematicians who seemed to proliferate in France at that time.

Fermat was an extraordinary genius. He was a lawyer by training, a member of the Toulouse parliament, and could devote to science only the spare time left from an extremely busy professional life. In mathematical circles, he is famous for a statement he wrote in the margin of a Greek treatise, adding that since he was short of space, he would write the proof elsewhere. That statement became known as Fermat's "great" or "last" theorem, and was proved only in 1993 by Andrew Wiles. Three centuries of progress in mathematics had been needed to bring it within reach. We have no idea what kind of proof Fermat had—or, much more likely, thought he had—but his intuition was right. So it was in the matter of the speed of light. Already in 1637 he had taken issue with Descartes, and he became more critical as the years went by. In 1662, he writes, "M. Descartes never proved his principle, for comparisons cannot serve as basis for proofs; in addition he makes a poor use of those he makes, and even assumes the passage of light to be easier through dense bodies than in lighter ones, which seems to be wrong."

Reflection of light is a much easier problem than refraction, and has been understood since antiquity: when a ray of light hits a reflecting surface, a mirror say, it bounces back at an equal angle (but at the other side of the vertical). In 1657, seven years after Descartes' death, Fermat received a treatise *On Light*, by one Marin Cureau de la Chambre, in which the law of reflection is stated and derived from a general principle, according to which "nature will always take the shortest way to act," meaning that light will travel by the shortest possible path between two given points. Cureau's argument is not original (nor does he claim it is): it goes back to the scientists and engineers in the first or second century AD, and can be found in the mass of writings attributed to one Hero of Alexandria. It is a beautiful argument, based on symmetry, and we reproduce it here for the reader to appreciate.

Fermat wrote back to Cureau, thanking him for the book. In his letter, he states his agreement with Hero's general principle, and raises a new question: "Since it has been useful for studying reflection, would it not be useful as well for refraction?" At first glance, it does not seem to be the case: to travel from point A to point B, the shortest path always is a straight line, but it is not the path along which light will travel if A lies in air and B in

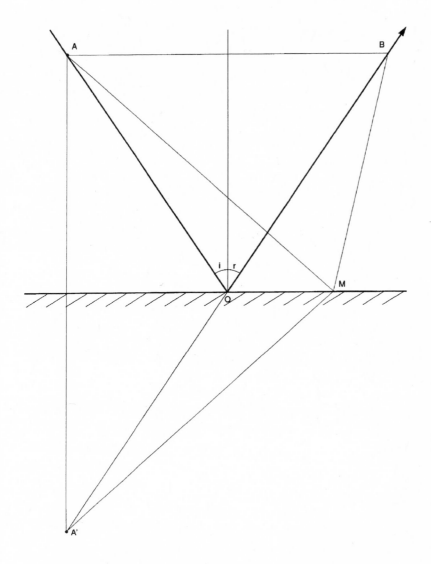

5. REFLECTION When a ray of light is reflected by a mirror, the incoming angle *i* is equal to the outgoing angle *r*. Hero of Alexandria noticed that this implies a remarkable physical property: if light travels from *A* to *B*, then it takes the shortest path between *A* and *B*. He even gave a mathematical proof. Take any two points *A* (on the incoming ray) and *B* (on the outgoing one). Denote by *A′* the point which lies in a symmetric position to *A* with respect to the mirror. Then the two paths *AOB* and *A′OB* have equal length, and so do the paths *AMB* and *A′MB*, where *M* is any other point on the mirror. Since *A′OB* is a straight line, it must be shorter that *A′MB*. So *AOB* must be shorter than *AMB*. This proves that the impact point *O* is positioned so that the path *AOB* is the shortest possible.

water. However, writes Fermat, if one accepts the idea that light travels faster in air than in water, the direct path from *A* to *B*, meaning the straight line *AMB*, will not be the fastest. As the picture shows, moving the crossing point *M* slightly closer to *O*, to *M'* say, will increase the distance traveled in air, but decrease the distance traveled in water. To be sure, the total length traveled, in air and in water, will be increased, as Hero's argument tells us, so that the time spent in air will increase and time spent in water will decrease; but since the speeds are not the same, the time lost in air will not be equal to the time gained in water, and it could well be the case that the time lost is less than the time gained. On balance, the broken line *AM'B* will be traveled faster than the straight line *AMB*. In fact, light behaves like a hiker traveling across a shifting landscape; if *B* lies in very difficult terrain, where progress is slow, it is the best idea to stay

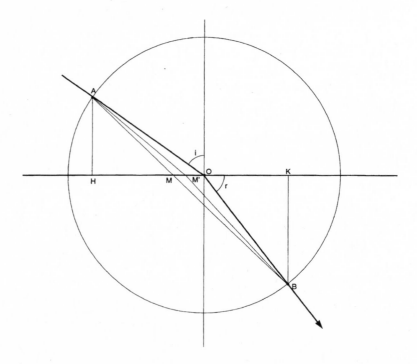

6. REFRACTION The horizontal line separates air (above) from water (below). The shortest path from *A* to *B* is the straight line *AMB*, and it would also be the quickest if light traveled as fast in air as in water. Since light travels faster in air, the path *AM'B* is actually quicker. It takes more time to go from *A* to *O*, and less to go from *O* to *B*; on balance, *AM'B* is faster than *AMB*, even though it is longer. Fermat showed that the quickest possible path from *A* to *B* is *AOB*, where the incoming and outgoing angles *i* and *r* satisfy Snell's law sin *i* = *n* sin *r*.

within the easier terrain as long as possible before crossing over. For instance, the hiker could aim for K, which is the point in the easy region which lies closest to B, thereby minimizing the amount of time spent in the difficult region. It may well be, however, that K lies too far from the starting point A, so that the hiker would gain time by aiming for some intermediate point H'. The best compromise is O, and that is precisely what Fermat is saying.

Fermat's letter to Cureau is quite remarkable. It is one of the earliest successes of mathematical modeling. There is first a general statement of a physical principle: to get from one point to another, light travels by the fastest (not necessarily the shortest) path. This principle is then applied to a new situation, namely refraction, yielding a mathematical problem: given two points A and B, separated by a line S, and a number n, find a point M on the line S such that the length $AM + nMB$ is the smallest possible. The connection between the physical problem of refraction and the mathematical model is through the number n, which tells us how much faster light travels in air than in water, so that Fermat is actually looking for the fastest path from A to B.

Although Fermat's letter to Cureau states the problem and goes very far toward the solution, it does not actually solve it. Fermat states that the solution of the problem will yield precisely Descartes' or Snell's sine law, $\sin i = n \sin r$, and he boldly announces that he will provide a mathematical proof: "I promise you in advance that I will find the solution whenever you like, and that I will draw consequences that will solidly establish the truth of our opinion. I will deduce at first: that the perpendicular ray does not break; that light breaks at the interface, without changing direction later on; that the broken ray moves closer to the perpendicular if it crosses from a rarer medium to a denser one, and moves away from it in the opposite case. In a nutshell, I will show that this opinion accounts exactly for all appearances."

Five years later, in 1662, Fermat delivered; he sent the solution to Cureau, in a letter brimming with enthusiasm, where the reader gets a rare glimpse into a mathematician's hour of triumph: "The price for my work turned out to be the most extraordinary, the least expected, and the happiest there ever was. Indeed, after running through all the equations, multiplications, antitheses, and the other operations which my method requires, and having finally solved the problem, as you will see in the enclosed sheet, I found that my principle yielded the very same proportion that Mr. Descartes had discovered for refractions. I was so taken away by such an unexpected result that I almost could not overcome my surprise. I have done my algebraic calculations over and over again, and the outcome has always been the same, even though my proof assumes that light travels faster through rarer mediums that denser ones, which I believe to be very true and necessary, although Mr. Descartes assumes the opposite."

A remarkable situation, indeed: two of the greatest mathematicians of all times, starting from assumptions which are in direct contradiction, end up at the same result. Descartes assumed light to travel faster in water that in air, Fermat assumed the opposite. They both agreed on the value of the refraction index n, namely 1.33 for the water/air interface, but they did not agree on its meaning: for Descartes, that number meant that light travels 1.33 times faster in water than in air, and for Fermat it meant that it travels that much slower. One could not disagree more: only one of them can be right. The controversy quickly became a fight. The Cartesians rallied to the defense of their dead master, and in their correspondence with Fermat sharpened blades showed under the flowers of courtesy. Their task was difficult: now that Descartes was dead, Fermat was acknowledged to be the best mathematician of his time, and there was no question that he had solved the problem he set himself: given two points in two mediums separated by a plane, given that one can travel in the first medium n times faster than in the second, find the quickest (and not the shortest) path between them. The solution is indeed given by the sine law, $\sin i = n \sin r$. What is open to attack is the relevance of Fermat's problem, which is purely mathematical, to the study of refraction, which is a physical phenomenon. One can understand that a weary traveler would try to figure out the quickest way home, and perhaps even resort to algebraic computations, but what about light? It has neither consciousness nor purpose, little does it care how fast it reaches a particular point, and there is no reason why it should prefer the quickest path, even if it knew it from the others. What is the basis for Fermat's claim?

In May 1662, Fermat received two letters from that same Clerselier who was to publish Descartes' *Treatise of the World* fifteen years later, and who was already a leader among the Cartesians. In these letters, he stated a certain number of objections to Fermat's approach, among which was the point we have just raised. Clerselier was not the most felicitous writer, and he put it as follows:

> The principle upon which you build your proof, namely that nature always acts by the shortest and simplest ways, is but a moral principle, not a physical one, which is not and cannot be the cause of any effect of nature. It is not, for it is not by this principle that it acts, but by the secret force and

virtue which lies in every thing; the latter not being determined by that principle, but by the force that lies in all causes that concur to a single action,
and by the disposition which is found in all bodies on which that force acts.
And it cannot be, otherwise we would be assuming some kind of awareness
in nature; and by nature, we mean here only that order and that law which
are established in the world as it is, and act without forethought, without
choice, and by a necessary determination.

There is no awareness in nature, says Clerselier. Attributing to nature
any sense of purpose, suggesting for instance that it is striving to minimize
some transition time, is not a scientific explanation, and any conclusion
that can be derived from this kind of reasoning must be dismissed.
Nature acts "without forethought, without choice," it does not look
ahead and it is never faced with choices. It does not pick its way among
several possibilities, taking into account their consequences, far or near
into the future; at any time, it finds just one door open, and it goes
through that door. This is what Clerselier means by "a necessary determination": since there will never be two open doors to choose from, the full
path is determined as you walk through the first door. The full story is
already written, you cannot change it, all you can do is to watch it unfold;
if you want to know more about the future, you need more information
about the present.

Nowadays, this view of the world is called *determinism,* and Clerselier
came very close to having coined the word. Determinism would be much
strengthened later on by Newton's discoveries, and would become the
prevailing view among scientists until the advent of quantum physics in
the beginning of the twentieth century. With quantum physics, nature is
sometimes given a choice and settles it randomly: whenever it is con-
fronted with several possibilities, it draws lots between them. Even today,
this seems a strange idea, and we feel much more comfortable with the
deterministic view of the world, which Einstein, for one, never aban-
doned: as he put it, "God does not play dice."

The argument against Fermat is that nature does not think ahead and
does not make decisions. Clerselier did not know Newton's laws, and his
idea of good science was modeled by Descartes' physics, which was based
on the idea that bodies interact only by direct contact: everything proceeded from the laws of collisions, as bodies great and small endlessly
bounced against each other. This is why so much of the scientific effort at
that time was devoted to studying what happens when two bodies or
more collide, with the aim of determining the outgoing velocities from
the incoming ones. Between collisions, the trajectories are just straight

lines; this was Galileo's principle of inertia, which Clerselier called to the rescue to challenge Fermat: "That way which you deem the shortest because it is the quickest is but a way of getting mistaken and lost, which nature does not follow nor would want to follow. For, as it is determined in everything it does, all it ever tends to do is to proceed in a straight line."

Here comes another argument against Fermat: we already know Galileo's principle of inertia, which satisfies the basic requirements of determinism (no looking forward, no open choices). Deriving it from other principles, such as the idea that light always travels the fastest path, is redundant, and so should be shorn by Ockham's razor. This second argument was taken up at later stages to dismiss Maupertuis' least action principle, just as Clerselier himself used it to dismiss Fermat's minimum time principle. Ernst Mach, for instance, in his great history of mechanics, published in 1883, stated that "the least action principle, and with it, all the minimum principles that one encounters in mechanics, simply express that, in every case, whatever happens is precisely what can happen under the circumstances, that is, whatever the circumstances determine, and determine uniquely." Farther on, he developed the idea to the conclusion that there is nothing more in Fermat's principle, or in his farther-reaching generalizations, such as Maupertuis' principle of least action, than the general fact that all phenomena of nature are fully determined by the relevant circumstances at the time of their occurrence. Neither Clerselier nor Mach gave a convincing argument for this downgrading of Fermat's and Maupertuis' principles, and in fact they were both wrong, as we will presently show: minimum principles do not follow logically from a general statement that the laws of nature are deterministic. They contain another type of information about the world.

Of course, in his answer to Clerselier, Fermat could not address issues which would be raised two and a half centuries later; but his letter, strangely enough, conveys some of the flavor of the famous controversy between Niels Bohr and Albert Einstein about the foundations of quantum mechanics. Here we go:

> Going back to the main question, it seems to me that I have often said, both to Mr. de la Chambre and to you, that I do not claim to be in Nature's secret confidence, nor have I ever claimed to be. It has obscure and hidden ways, which I have never tried to penetrate; I had only offered it some slight geometrical help in the matter of refraction, in case it had needed it. But since you assure me, Sir, that it can proceed to its business without that, and

that it is content with following the way Mr. Descartes has prescribed, I heartily abandon you my pretended conquest in physics, provided you leave me in possession of my problem in geometry, all pure and *in abstracto*, whereby one can find the path of a moving object which crosses two different mediums, and which tries to end its motion as soon as possible."[7]

Fermat associated a mathematical problem (the model) with a physical phenomenon (refraction). Clerselier objected that there is no reasonable meaning to be attached to the model: things cannot actually work that way, it cannot be that light has both the desire to travel fast and the means to compute the quickest path. Fermat answered that light propagates *as if* it had both that desire and these means, and while the mathematical problem may not be an accurate description of what is happening at some deeper level of reality, it is good enough to make predictions which turn out to be in agreement with experiments. So the model should be kept as a working tool for scientists, until it is discarded for a better one, and the question of why it works and what it means should be left to philosophers to worry about.

This was a very modern position, precisely the one Bohr would take against Einstein: do not worry about the meaning of the mathematical model, as long as it is logically coherent and it accounts for observations. Einstein claimed that God does not play dice; Bohr answered: "I don't know; all I am saying is that, using quantum mechanics and probability theory, I can make very accurate predictions." Clerselier claimed that nature cannot show purpose; Fermat answered: "I don't know; all I am saying is that, using a minimum principle and some calculus, I can account for the refraction of light." Fermat and Bohr could not be farther from Leibniz, who believed that God had created the world with a definite purpose in mind, namely to make it as perfect as possible; that purpose must then lie at the heart of all physical laws, and must be their hidden meaning. One might even think that all of physics could be recovered from that single idea, and that a real scientist should aim for that inner core of reality; this is what Maupertuis will claim to achieve in the following century. But Fermat's position was already firm: science does not need this. It would have been fascinating to watch that controversy develop. Unfortunately, Fermat died three years later, in 1665. The Cartesians faced a much more formidable adversary: Isaac Newton.

7. P. Tannery and C. Henry, eds., *Oeuvres de Pierre de Fermat* (Paris: Gauthier-Villars et fils, 1891–1894).

Newton's ideas did not meet with immediate success, even in Great Britain; and France, under the influence of Descartes' disciples, remained a stronghold of resistance for fifty years. In France, the fight against Cartesian physics was part of the fight against Cartesian philosophy, and more generally the old order of things, the ancien régime which would finally be brought down by the 1789 revolution. The main leader in this fight was Voltaire, whose astonishing activity extended over all aspects of intellectual life. In 1733, he published, in English and in French, twenty-four *Letters Concerning the English Nation*, which contained an enthusiastic account of Newtonian physics. His companion of many years, the marquise du Châtelet, wrote an excellent French translation of the *Principia*, to which Voltaire contributed an introduction in verse.

In the Paris Academy of Sciences, the fight quickly crystallized around a specific question, namely the shape of the Earth. It had been known to be round since one of Magellan's ships had sailed westward from Spain to Spain between 1520 and 1522. But it was not a perfect sphere. Newton, working on the idea that the Earth is a liquid ball that had solidified, had predicted that it would be flattened at the poles, because its rotation when it was fluid would have created a bulge around the equator. Cassini, the French astronomer royal, a loyal Cartesian, believed the opposite: the Earth should be elongated at the poles, somewhat like a lemon. Measuring two arcs of meridian, one near the pole and one near the equator, would settle the question. Indeed, an arc of meridian is the distance between two points on the surface of the Earth which would be seen from the center at an angle of exactly one degree. If the Earth were a perfect sphere, this distance would be the same all over the globe, and would be equal to $P/360$, where P is the circumference of the sphere. If it were not a perfect sphere, this distance would depend on where it is measured: if Newton is correct, it must be smaller near the pole than near the equator, if Cassini is correct, it must be greater.

In time, this question came to be seen as a litmus test between Cartesian and Newtonian physics, so that the Academy decided to send two expeditions to measure the arc of meridian, one near the North Pole, the other near the equator. In 1736 the two expeditions left, one for Peru, the other one for Lapland. The latter was placed under the leadership of a mathematician aged thirty-four, named Pierre Moreau de Maupertuis, who in 1732 had written a brilliant essay on gravitation. He was one of the most remarkable characters of a time with no shortage of extraordinary figures. The Peruvian expedition took ten years, but Maupertuis was already back by 1737, sixteen months after he had left. The measurements

7. PIERRE MOREAU DE MAUPERTUIS (1698–1759) In this portrait, Maupertuis is represented flattening the Earth with his right hand, a reference to his famous northern expedition of 1736–1737. There is also, at the bottom, a reindeer pulling a sleigh, one of the many adventures that Maupertuis described vividly on his return to France.

he brought, compared with the arc of meridian at the latitude of Paris, showed that Newton was right, and made Maupertuis a hero overnight.

Father Outhier, who accompanied him, wrote an account of the expedition. He tells of the many difficulties that met them in the northern lands, of being devoured by mosquitoes and flies, of gliding on the snow with strange planks strapped to their feet, of pushing themselves forward with sticks, of continually falling down and not being able to get back on their feet. His book carries an illustration of "a Lapp walking on the snow with one pine plank at each foot, and a stick with a circle at the end so as not to sink into the snow." True to this description, Maupertuis brought back with him the first pair of skis that had ever been seen in France. Two native Lapp girls followed him, and they were a great success in Paris, where they eventually found spouses and settled . Maupertuis became immensely popular, and in 1745, Frederick the Great, the philosopher-king of Prussia, called him to Berlin to preside over the newly founded Academy of Sciences. He held that position until his death in 1759, carrying out scientific activities on a great many different subjects, from mathematics and physics to biology, as we shall see. We should note, for instance, that Maupertuis holds the honor of being the first scientist ever to have stated the idea that animal and plant species are not immutable. In his books, strangely titled *The Physical Venus*, and *The White Negro*, he puts forward the idea that populations can evolve because of external circumstances and the accumulation of small changes over long periods of time. Of course, he brings no serious evidence in support of these revolutionary ideas, but there is no lack of merit in putting them forward at a time when everyone else thought that elephants had been around since the beginning of the world.

Maupertuis' was an eventful life, full of impressive achievements. He certainly left his mark on the world, and should be recorded as one of the leading figures of the French Enlightenment. Unfortunately, in his Parisian days, he fell afoul of Voltaire. The Parisian literary circles, the famous salons where wits were perpetually matched against each other and a sharp tongue was prized above everything else, were not the proper place for friendships to develop. All the leading intellectuals of that time were at odds with each other, Voltaire against Rousseau, d'Alembert against Diderot; but Maupertuis gained Voltaire's special enmity, subdued at times, but always ready to surface. When Maupertuis came back from his northern expedition, and all of Paris sang his praises, Voltaire chimed in with these verses:

> You have gone to confirm, in places far and lonesome
> What Newton always knew without leaving his desk.

True genius stays at home, while lesser men run to Lapland, hardly a laudatory account. Later on, things would become much worse. Voltaire had been corresponding with Frederick II of Prussia for a long time, and finally he accepted the king's invitation to join his court in Potsdam. Their relationship soured, and a few years later Voltaire fled Prussia in humiliating circumstances, bringing with him an undying hatred of Maupertuis, who, as president of the Academy in Berlin, stood for all of Frederick's intellectual pretences. From then on, ridiculing Maupertuis was a way of getting even with the king, and no opportunity to do it was ever missed. For instance, in 1753, during the Seven Years' War, Maupertuis was captured by the Austrians, who at the time were fighting the Prussians, sent to Vienna, and then set free because of his renown as a scientist—hardly a dishonorable episode. Voltaire relates it as follows: "He has been captured by some Moravian peasants, who stripped him naked and emptied his pockets of more than fifty theorems."

Enough of this for the moment. Let us go back to Maupertuis himself. Among his many scientific interests, Newtonian mechanics had always been prominent. In 1732, for instance, he had published a scientific memoir on the law of gravity, which had certainly been instrumental in his designation as head of the Lapland expedition. In 1744, just before leaving for Berlin, he published in Paris another memoir entitled *Agreement between Several Laws of Nature Which until Now Had Seemed Incompatible*. Under that pompous title, Maupertuis, after Descartes and Fermat, reopened the case of refraction. He dismissed both his predecessors, the first for being guilty of having likened light rays to material balls, the second for having assumed that light travels faster in air than in water, and he puts forward his own explanation: "After deeply pondering this matter, I have reached the conclusion that light, when it crosses from one medium to another, since it already abandons the shortest path, which is the straight line, may as well abandon the quickest one: for what precedence should there be between time and space? Since light cannot travel simultaneously the shortest way and the quickest one, why should it choose one rather than the other? So it follows neither; it takes a road which has a more real advantage: the path it travels is the one for which the quantity of action is least."

Note Maupertuis' exasperating arrogance, which did little to make him popular among his peers. He goes on to explain what he means by *action*, or rather *quantity of action*. There is a real difficulty here, because that word has a familiar meaning that immediately jumps to mind, but which has nothing to do with the technical meaning Maupertuis has in mind: "When a body is carried from one point to another, some action is needed: this action depends on the velocity with which the body is moving, and the distance it is traveling, but it is neither the one nor the other, taken separately. The greater the velocity and the longer the distance, the larger the quantity of action: it is proportional to the sum of distances traveled, each of which is multiplied by the velocity with which it is traveled."

In other words, if a body travels from point A to point B in a straight line, with constant speed v, the quantity of action for that motion will be the product mvl, where m is the mass of the body, v its velocity, and l the distance between A and B. If the path from A to B is no longer straight, but broken, consisting of straight portions traveled with constant speeds, the quantity of action for each portion much be computed according to the preceding formula, and they must all be added up to get the total quantity of action associated with the given motion from A to B. Maupertuis then shows that the sine law for refraction follows from this formula: taking the refraction index n to mean that light travels n times faster in water than in air, and looking for the path of least action connecting point A in air with point B in water, leads to the formula $\sin i = n \sin r$. Maupertuis concludes that "this quantity of action is what Nature is truly spending, and what it is trying to save as much as possible during light's travel."

For all of Maupertuis' crowing, Fermat was the one who was right: light travels faster in air than in water. If Maupertuis gets the sine law, although he made the wrong assumption, it is because he made an earlier mistake which, instead of compounding the second one, corrects it. In other words, his physics were doubly wrong, and he was lucky enough that it yielded the correct mathematical answer. In the tradition of Descartes, Maupertuis considers light to consist of massive particles, which accelerate as they enter water. But if their speed changes, so must their energy. A century later, Carl Jacobi would show that Maupertuis' least action principle holds only in cases where the energy does not change during the motion (so-called conservative systems), so that the application Maupertuis made of his own principle to the refraction of light was illegitimate. That the end result was correct was just a fluke.

There would be no more to say about the 1744 memoir, if it had not given its author the idea of a general minimum principle, applicable not only to light, but to all the problems of mechanics. In fact, Maupertuis' closing words are that "all phenomena of refraction now agree with the great principle that Nature, to produce its effects, always acts by the simplest ways." The next year, Maupertuis was in Berlin, and published a new memoir entitled *The Laws of Motion and Rest Deduced from a Metaphysical Principle*. This, in Maupertuis' words, was the general principle that "the quantity of action necessary to cause any change in Nature, always is the smallest possible," and was henceforth known as the *least action principle*. As an example, Maupertuis derived from this principle the motion of two bodies undergoing an elastic shock. Simultaneously, the great Leonhard Euler published in Latin a book entitled *A Method for Finding Curves Which Are Maximizing or Minimizing*, with an appendix where he derived even more consequences of the least action principle.

Maupertuis himself, leaving mathematics to better hands, waxed metaphysical and probed the deeper meaning of his discovery. In 1752, he published *Essay in Cosmology*, where he stated, with all due modesty, "After so many great men who have worked on this matter, I am almost afraid to state that I have discovered a principle underlying all the laws of motion, which applies to hard bodies as well as elastic ones, from which all motions of all corporeal substances depend.... Our principle, more in conformity with the ideas we should entertain about things, leaves the world in its natural need of the Creator's power, and follows naturally from the use of that power. ... How satisfying for the human spirit to contemplate these laws, so beautiful and simple, which may be the only ones that the Creator and Ordainer of things has established in matter to sustain all phenomena of this visible world." As he saw it, the least action principle was God's mark on his creation, and it fell to him, Maupertuis, to discover it by purely scientific means. God's operation in nature was now clear to the human eye: he always acts to spend as little as possible of that mysterious quantity mlv. This was indubitable proof of a divine purpose, and hence of the existence of a creator. Whoever finds that the laws of physics all tend to that single purpose of consuming as little as possible of that mathematical fuel will have to agree that these laws must be due not to chance, but to design.

And what design could God have upon the world except to make it better? The actual world, ruled by the principle of least action, must be the best possible, and so it must be that the quantity of action somehow expresses the amount of good (or bad, rather, since God wants to make it as small as possible). In Maupertuis' words, still from his 1752 book, "Once it becomes known that the laws of motion are founded on the principle of the *better*, no one will doubt that they are due to an *all-powerful and all-wise Being*, who may have given bodies the power to act upon each other, or who may have used some other way which is even less known to us." In fact, he claims a grand unification of his own, the unification of physics with metaphysics, and even with morals. In later work, he claims that a certain quantity of good (or bad) is attached to each of our actions, and that God has ordained the world so that, adding up the good and subtracting the bad, the balance will be found to be the greatest possible.[8] In other words, this is the best of all possible worlds. From then on, thanks to Voltaire's talent, Maupertuis is remembered as Doctor Pangloss, the incurable optimist of the novel *Candide*, who manages to find in minuscule consequences of major catastrophes evidence to reassert his steadfast belief that the good always outweighs the bad.

This took some doing on Voltaire's part. Maupertuis enjoyed a strong position, as president of the Berlin Academy of Sciences, and knew how to use it. The occasion was provided when Maupertuis got involved in a priority quarrel. In March 1751, a professor in the Netherlands named

8. *Essay on Moral Philosophy* (1741).

Koenig, an old acquaintance of Maupertuis, published in the prestigious journal *Acta Eruditorum* a review of the least action principle, where he referred to a letter written by Leibniz in 1707. A copy of the letter was provided as an annex to the paper, and it said the following: "The action is not as you had thought, time should enter into it, it is as the product of mass by distance and time. I have noticed that, in the modifications due to motions, it usually becomes a maximum or a minimum. One could derive from that several statements of great import; it could help to determine the trajectories of bodies attracted by one or several others."

This looked like evidence that Leibniz had discovered the least action principle before Maupertuis. In itself it does not diminish Maupertuis' merit, the more so as Leibniz's letter (or rather, Koenig's copy) goes on to explain that he has given up working on dynamics, because his views have not been accepted. It is undoubtedly Maupertuis who introduced the least action principle to the scientific community, and who did the mathematical work Leibniz had barely sketched. Scientific discovery does not consist of stating ideas and leaving others to check whether they are correct or not. The gravitation law, for instance, is attributed to Newton, but he was not the first one to have stated it: Robert Hooke, for one, had done it before him. On the other hand, Newton certainly is the first one to have shown that all three Keplerian laws are mathematical consequences. There is no reference to Hooke in the *Principia*, but even if Newton had picked the idea of the inverse square law from someone else, there were very many other ideas lying around from which to pick. Newton's merit lies in choosing the right one, and in the wealth of consequences he derived from it with extraordinary mathematical skill and physical insight.

Since Leibniz had not developed the idea of a least action principle, and had not even published it, Maupertuis could have let matters lie, the more so as Koenig's review was far from aggressive. Instead, he unwisely chose to accuse Koenig of forgery, and to bring the Berlin Academy to the rescue. Challenged to produce the original of Leibniz's letter, Koenig claims to have seen it at the home of a friend of his named Henzi, who unfortunately turned out to have been beheaded in Bern in 1749. No letter of Leibniz was found in Henzi's papers, so on April 13, 1753, the Academy voted that "this fragment has been forged, either to disparage Monsieur de Maupertuis, or to exaggerate by a pious fraud the praise due to the great Leibniz." It should be noted at this stage that another copy of Leibniz's letter was discovered in 1913, so that today there is little doubt as to its authenticity. No wonder that Koenig rose to the challenge: in September of the same year, he set forth his *Appeal to the Public*. After that, Voltaire joined the fray. His book, *Story of Doctor Akakia and the Native of Saint-Malo* is a collection of pamphlets against Maupertuis, the general theme being that such a mass of nonsense had been published in recent years under the name of the respected president of the Berlin Academy of Sciences that it is simply not possible that they were authentic: they had to be the work of a young impersonator, whom Voltaire proceeds to unmask.

The opening lines set the tone: "The native of Saint-Malo[9] had long fallen a prey to a chronic sickness, which some call philotimia[10] and others philocratia.[11] It went so strongly to his brain, and he had such strokes, that he wrote against medicine and against the proofs of God's existence. Sometimes he imagined himself digging a hole to the center of the Earth, other times building a Latin town. He even had some revelations about the workings of the soul by dissecting monkeys. He finally got to the point where he thought himself to be greater than a certain giant of the preceding century, named Leibniz, although he was not quite five feet tall."[12] All these, of course, allude to ideas or experiments of Maupertuis, taken out of context. Later on, the impersonator is tried by a board of professors of wisdom, who rule that "it appears that this young author has taken but half of Leibniz's idea; be it known that he never held a whole idea of Leibniz"—doubtless from lack of capacity. There is also a "memorable sitting" of an unnamed academy, where the president tries to support his ideas on biology by coupling a mule and a peacock, and germinating wheat spontaneously turns into fish terrines to feed the ladies. At the end, the young impersonator implores his pardon: "We ask God to forgive us for having claimed that there is no other proof of his existence than A plus B divided by Z, and we beg the gentlemen inquisitors not to judge us too harshly in this matter, which they do not understand any better than we do."[13] So much for Maupertuis' view that the principle of least action shows God's hand at work in nature.

Voltaire's pamphlet was an instant success, and Maupertuis was ridiculed throughout Europe. He died in 1759 in Basel, a broken man. The final blow came after his death: it is *Candide*, Voltaire's masterpiece, a satire of philosophical optimism which is read even today. Maupertuis is reborn as Doctor Pangloss, who professes that "all is well that ends well in the best of all possible worlds." Disasters continuously befall him.

9. Maupertuis was born in Saint-Malo, on the northern coast of Brittany.
10. In Greek, love of honors.
11. In Greek, love of power.
12. Voltaire, *Histoire du docteur Akakia et du natif de Saint-Malo* (Paris: A. G. Nizet, 1967).
13. Ibid.

The beautiful castle where the Baron de Thunder-ten-tronckh entertains him as a professor of philosophy is destroyed and his benefactors killed. He wanders through Europe and South America, where he witnesses the horrors of war and of slavery. He is in Lisbon on November 1, 1755, when an earthquake destroys the city and kills forty thousand. Nothing can cure Pangloss of his incurable optimism. At the end of the novel, meditating in his garden, he concludes that if all this had not happened, he would not be sitting there in the shade eating pistachios. Hardly a fair assessment of Maupertuis' science and philosophy.

In scientific circles, however, life went on, and Maupertuis' principle was subject to scrutiny. It states that, among all possible motions, nature picks the one with the least quantity of action. This seems like a simple statement, but it is not. It raises several questions. The first is how to define precisely the quantity of action, and we have seen how Maupertuis answered it. But there are others: what is meant by a "possible" motion? Since they do not occur, how are we to compare them with the actual motion, the only one we can observe? What would be meant by an "impossible" motion? In fact, the situation is so tricky that one century later, Carl Jacobi (1804–1851), in his celebrated *Lectures on Dynamics*, would declare that "this principle is stated in all treatises, and even in the best ones, those by Poisson, Lagrange, and Laplace, in such a fashion that, as far as I am concerned, it cannot be understood." It fell to Leonhard Euler (1707–1783), Joseph-Louis Lagrange (1736–1813), William Rowan Hamilton (1805–1865), and finally Jacobi himself, to formulate Maupertuis' ideas in a precise and workable way.

The great Euler, prince of mathematicians, heads this list. In the appendix to his book of 1744, he applied the least action principle to several interesting examples, such as the free fall of a heavy body, or the motion of a body undergoing attraction to a fixed center. For that purpose, he introduced new ideas and methods into mathematics, thereby creating a new field, called the *calculus of variations*, which has been extremely active ever since. In addition, Euler extended Maupertuis' definition of the quantity of action, which was restricted to linear motions with constant speed, to much more general situations, where the body moves along a curve and the velocity changes with time. With Euler's definition, the quantity of action can be computed for any kind of motion arising in Galilean mechanics or Newtonian physics. In fact, Euler's definition is so comprehensive, and the consequences he derives are so striking, that he might well have been considered the discoverer of the least action principle, if he had not wisely declined that honor. During the controversy with Koenig, Euler sided with Maupertuis; in 1753, he

·published *Dissertation on the Least Action Principle*, where he refuted Koenig's criticisms and unequivocally gave the priority to Maupertuis: "I will not comment here on the observation I have made that in the motion of heavenly bodies, and more generally in the motion of all bodies attracted by a center, if at any time one multiplies the mass by the distance traveled and by the speed, the sum of all these products always is the smallest possible. Since this observation appeared in print only after *M.* de Maupertuis had exposed his principle, it cannot detract from its novelty."

"Euler, a truly great man, left the least action principle with the name it already had, and Maupertuis the glory of its discovery, but he turned it into something new, practical and useful." This is how Mach, in his *History of Mechanics*, summarized Euler's contribution.[14] In 1754, a young mathematician named Lagrange, inspired by Euler's work, discovered a general method for solving problems in the calculus of variations. Central to his approach was a system of equations, now called the Euler-Lagrange equations, and in 1756, he showed how all of Galilean mechanics can be derived from the least action principle by a simple application of his general method. In Lagrange's words, a *principle* is "a simple and general method for solving all imaginable problems in dynamics, or at least to write the corresponding equations."[15] If such a principle is not known, "one would always need a particular sleight-of-hand to unravel, in each problem, the forces which should enter into consideration, which made these problems exciting and competitive." On the other hand, once such a principle is known, the least action principle, for instance, it no longer requires creativity and ingenuity to solve problems: all one has to do is to apply the standard method, to wit, the general principle. The excitement and competitiveness may be gone from the field, but it now is open to all, and problems can be solved much more efficiently; researchers have given way to engineers. After Lagrange, solving problems in mechanics, that is, finding the equations describing the motion of rigid bodies or systems of particles, will no longer require the genius of Galileo or Fermat; it will simply be a matter of having learned and understood the Euler-Lagrange equations. This is the essence of scientific progress.

As Mach puts it, "Science itself can be considered as a minimum problem, consisting in accounting for facts as perfectly as possible, at the

---

14. Ernst Mach, *Die Mechanik in ihrer Entwicklung historisch-kritisch dargestellt* (Leipzig: Brockhaus, 1883); English translation by Thomas J. McCormick, *The Science of Mechanics: A Critical and Historical Exposition of Its Principles* (Open Court, 1893).

15. *Analytical Mechanics* (1788), 179.

smallest intellectual expense." Lagrange summarizes half a century of work in his epoch-making treatise, *Analytical Mechanics*, first published in 1788, and in the introduction he proudly declares: "No pictures are to be found in this book. The methods I explain need neither constructions, nor arguments from geometry or mechanics, but only algebraic operations, carried out in an orderly and uniform fashion. All who like calculus will enjoy seeing mechanics becoming one more branch of that discipline, and will be grateful to me for having extended its domain." With this declaration, one can see what degree of maturity had been reached by mechanics: there was no more need to imagine the physical system one was studying; one could write down directly the equations of motion.

Lagrange's point of view, that mechanics is about writing down systems of equations and solving them, was dominant until the end of the nineteenth century, when Poincaré brought mechanics into geometry. Until then, Lagrange's *Analytical Mechanics* remained the basic reference, and its influence was pervasive in teaching and research. The first part of the book is devoted to explaining the four great principles of mechanics, among them the least action principle, which Lagrange states essentially as Euler did. However, for reasons we will go into later on, he will not use it in the rest of his treatise, preferring to rely on other principles of mechanics. As a result, he is rather sketchy in describing the least action principle, and leaves the reader with unresolved doubts and ambiguities. For instance, it is not obvious what the "possible" motions are which the "real" motion is to be compared with, or how the quantity of action is to be computed along these unreal motions. The least action principle states that, of all possible motions, only one will actually happen, the one that yields the least quantity of action, but giving a precise mathematical meaning to this statement is not as easy as it seems. In fact, one had to wait for Hamilton and Jacobi to get a fully satisfactory account.

The modern statement of the least action principle is formulated, not in the standard three- or two-dimensional space where motions take place, but in the so-called *phase space*, which is the main discovery of Hamilton. The basic idea is to record, at every instant, not only the position of the moving body or system under consideration, but also its velocity. The succession in time of positions and velocities defines a path which is no longer in the standard space, as it would be if only positions were taken into consideration, but in a phase space, with twice as many dimensions, and it is this path which should be considered when applying the least action principle. I shall give details and examples in later

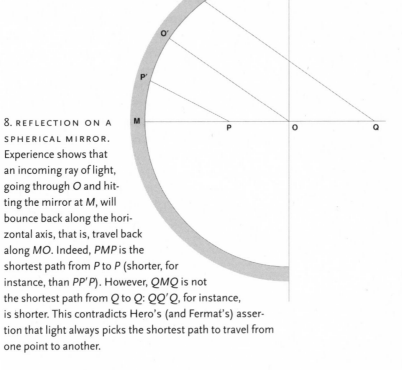

8. REFLECTION ON A SPHERICAL MIRROR. Experience shows that an incoming ray of light, going through *O* and hitting the mirror at *M*, will bounce back along the horizontal axis, that is, travel back along *MO*. Indeed, *PMP* is the shortest path from *P* to *P* (shorter, for instance, than *PP'P*). However, *QMQ* is not the shortest path from *Q* to *Q*: *QQ'Q*, for instance, is shorter. This contradicts Hero's (and Fermat's) assertion that light always picks the shortest path to travel from one point to another.

chapters; suffice it at this point to say that, with the introduction of phase space, Hamilton and Jacobi found the right mathematical setting for the least action principle. And they also found out something more, namely that this principle is misnamed: the quantity of action is not made as small as possible (minimized), or as large as possible (maximized); it is made *stationary*.

Early on, it had been pointed out that the quantity of action is not always minimized. In 1752, for instance, one Chevalier d'Arcy sent a memoir to the Paris Academy of Sciences, in which he studied the reflection of light on the inside of a spherical mirror. Let us denote by *P* the point where light is emitted, and *O* the center of the sphere. Chevalier d'Arcy shows that the least action principle holds only if *P* is closer to the mirror than *O*; if it lies farther away, the ray of light starting from *P* and hitting the mirror perpendicularly is not the shortest one from *P* back to

P. Strangely enough, not only Maupertuis, but Euler and Lagrange over-looked this example, and held throughout their lives the conviction that actual motions always minimize the quantity of action among possible ones. Hamilton was the first to analyze the situation correctly, and to state that the quantity of action is actually made stationary.

The concept of a stationary path really belongs to mathematics. It is like the sweet spot on a tennis racket: not apparent on inspection, but clear enough to the player. It means that, first, one will compare the actual motion, or the corresponding path in phase space, only to those paths which are close by, in fact as close as possible and even closer. Second, the quantity of action will be insensitive to changes in the underlying path: small changes in the motion will cause even smaller changes in the quantity of action. The situation is similar to a mountain pass separating two peaks and two valleys. Peaks correspond to maxima, points where the altitude is highest, but the passes correspond to stationary points. To see it as a mathematician, one should imagine a heavy fog lying over the scenery, so that one cannot see any farther than one's own feet; the mountain pass is recognizable by the fact that the ground is

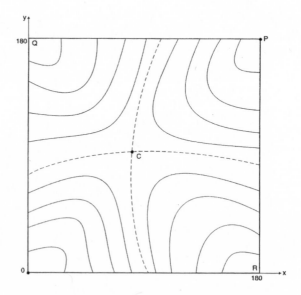

9. A STATIONARY POINT The solid picture represents the graph of a function of two variables *x* and *y*. There are two peaks (on the left and on the right), and between them a mountain pass separating two valleys (one in front of the pass and one behind it). This mountain pass is located at a stationary point *C* of the function. The other part of the picture shows the level curves (points of equal

horizontal. At every other point, there will be a definite slope, down which water will flow, but at a stationary point water will stay in unstable equilibrium, uncertain which way to go. A geometrical abstraction of that situation would be a saddle: the stationary point is the only point where one can sit vertically. It is neither a top (point of maximum height) nor a bottom (point of least height), but it is the only point where the surface of the saddle is horizontal. At every nonstationary point, the saddle is sloped in a certain direction. At a stationary point there is no slope: a marble put exactly on that spot would remain there.

Note that a peak or a lake would enjoy the same property: at the top of the peak, or at the bottom of the lake, the ground has to be horizontal. So maxima or minima are particular cases of stationary points; there are other kinds of stationary points than these two, as the preceding example shows. General stationary points are not as easy to visualize as maxima or minima. This is probably the reason why Maupertuis, Euler, and Lagrange overlooked them and did not question the fact that the laws of mechanics would minimize the quantity of action. In his *Lectures on Dynamics* taught in 1842–1843, Jacobi says that "when the principle of

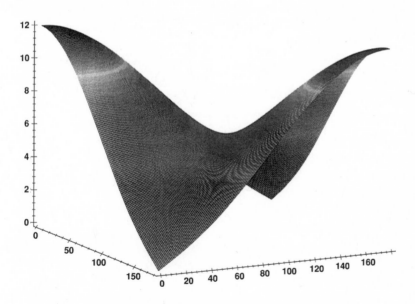

height), as they would appear on a geographical chart of the area. They rise toward the two peaks, located at Q and R, and sink toward the two valleys, located at O and P. At the stationary point C, two level curves cross. All the points on the dotted lines lie at the same altitude.

least action is stated ... it is usually asserted that [the quantity of action] must be minimum or maximum, instead of saying that it is stationary. The mistake is now so common, that it can hardly be held against those who make it." He then proceeds to give another example where the quantity of action is not minimized: it is the motion of a body gliding without friction (or gravity) on the surface of a sphere. If the principle of least action were true, the motion would always take the body from one point to another by the shortest path. But the motion is easy to figure out: the trajectory is a great circle on the sphere, and the moving body accumulates loops around that circle. Certainly, once it has gone several times around the sphere, it is no longer taking the shortest path: all the loops are extra, useless additions to the distance traveled.

It is a rare pleasure in mathematics to find mistakes of one's predecessors, especially if they have the caliber of Euler or Lagrange, and Jacobi really rubs it in:

> One even finds, concerning the matter of the shortest paths, a remarkable quid-pro-quo in Lagrange and Poisson. Lagrange very properly states that the quantity of action can never be a maximum, for, however long one may draw a curve on a given surface, it is always possible to draw a longer one; and he concludes that this quantity must always be minimum. Poisson, on the other hand, who knew that in certain cases,[16] when the motion occurs on a closed surface and the quantity of action becomes too large, its value ceases to be a minimum, concludes that in such cases it must be maximum. Both conclusions are wrong. In truth ... the quantity of action can never be a maximum; it can be either a minimum, or neither a maximum nor a minimum."[17]

One century after Maupertuis' first memoir, we finally have a fully correct and comprehensive statement of the least action principle. Unfortunately, mathematical precision has come at the expense of intuition. Paths have to be plotted in phase space, not in the two- or three-dimensional space where the motion actually takes place, and the quantity of action is evaluated by a complicated mathematical formula. More importantly, it is no longer a minimum: real motions do not always make the quantity of action as small as possible; they may merely make it stationary, which is another mathematical subtlety. We are far from Maupertuis' simplistic idea of God saving on the quantity of action he needs to run the world.

This idea had been shared by Euler. In his book of 1744, he had stated that "since the constitution of the universe is perfect, and completed by

16. The case of the sphere, which we just discussed.
17. *Lectures on Dynamics* (1866).

an all-wise creator, absolutely nothing happens in this world which cannot be explained by some argument of maximum or minimum; this is why there is no doubt at all that all effects observed in the world can be explained from final causes, with the method of maxima and minima, with the same success as from efficient causes."[18] What Euler is saying here is that one can actually teach mechanics and physics by starting from the idea that God wants to minimize the quantity of action spent in running the world and deriving its consequences. Another great mathematician of that time, who wrote an important *Treatise on Dynamics* (1758), Jean d'Alembert, took the opposite view:

> It seems to me that this should help us evaluate the proofs that several philosophers have given for the laws of motion and which rely on the final causes principle, that is, on the purpose that the Author of nature must have set himself when establishing these laws. Such proofs cannot carry conviction unless they are preceded by, and rely on, direct proofs drawn from principles which are closer to our understanding; otherwise, they would frequently lead us into error. It is because he followed that direction, because he believed it was part of the Creator's wisdom to keep constant the quantity of motion in the universe, that Descartes got the laws of collision wrong. Those who would imitate him would indeed run the risk [of] being mistaken as he was, or to state as a general principle something which happens only in certain cases, or finally to consider a fundamental law of Nature a mere mathematical consequence of a few formulas.

One of the basic survival rules for the young scientist is to avoid controversies. Lagrange wisely sidestepped the whole issue. In his *Analytical Mechanics*, he stated four different principles, each of which can serve as a foundation stone for mechanics, and among them the least action principle. But he chose another one as a basis for his own exposition. In his own words, "I regard [it], not as a metaphysical principle, but as a simple and general consequence of the laws of mechanics. One will see, in volume 2 of the Turin memoirs, how I used it to solve several difficult problems in dynamics. This principle, combined with the one of conservation of energy, and developed according to the rules of the calculus of variations, gives directly all the equations which are necessary to solve every problem, and hence a simple and general method for treating questions about the motion of bodies; but this method itself is nothing but an offshoot of the one I will describe in the second part of this work, and which

18. *Method to Find Curves Which Are Maximizing or Minimizing.*

has the added advantage of being derived from the first principles of mechanics."

Hamilton, as we have seen, made the point that, contrary to what Maupertuis, Euler, and even Lagrange had believed, actual motions do not minimize the quantity of action. This efficiently stripped the least action principle (now recognized to be misnamed) of all pretense to metaphysics. Whereas one could envision the Creator saving the fuel the universe is running on, it is very difficult to imagine him going to similar trouble just to keep the quantity of action stationary; it takes a mathematician to know a stationary path, and the comparative advantage of such paths over others is far from obvious. Hamilton rightly concludes that "though the least action principle has now taken place among the highest of physics, its pretensions to express a cosmological necessity based on the thrift of nature are now generally rejected. And this rejection seems justified by the simple argument—among others—that the quantity which is supposedly saved is in fact often spent with unbounded prodigality."[19] Though its metaphysical status was gone, the least action principle still kept, in Hamilton's eyes, a prominent position among the laws of mechanics; not so with Jacobi who, as usual, drove the nail into the coffin:

> One can find, in Euler's book *On the motion of projectiles*, which we quoted earlier, an example where this principle is used. After having established it in the case when there is a fixed attracting center, he cannot extend it to the case when there are two mutually attracting free bodies, because he did not know the conservation of energy; he is then content with saying that in this latter case the calculations would be very long, but that the principle of least action must remain valid anyway, for the foundations of a healthy metaphysics require that in nature, forces must realize the smallest possible action (according to him, because of gravity in the bodies). But there is no question here of a healthy metaphysics, nor even of metaphysics at all, and actually the only reason Euler could have written that is because he had been misled by the name of "least action."[20]

This spelled the end of the least action principle as a metaphysical tool. The clinching argument in this controversy was Hamilton's discovery that the quantity of action is not minimized, but made stationary.

19. "On a General Method of Expressing the Paths of Light, and of the Planets, by the Coefficients of a Characteristic Function," *Dublin University Review* 1833.
20. *Lectures on Dynamics.*

From then on, the least action principle became a purely mathematical tool, the usefulness of was not fully understood until the twentieth century. Before that, there was another challenge to overcome.

The first attack came from Jacobi, still in his *Lectures on Dynamics*. He stated that "the importance of this principle is due, first to the way in which it enables us to write the equations of motion, and, second, to the fact that it gives a function which becomes minimum if the equations of motion are satisfied. Such a minimum always exists, but in general it is not known where: in earlier times, an overblown importance was attached to the fact that such a minimum exists at all, whereas the true importance of the principle is that this minimum can be given *a priori*." It is not perfectly clear what Jacobi meant. Apparently he was saying that it is no miracle that there is a quantity which the motion will minimize; the real wonder is that one knows what it is, namely the quantity of action, so that one can write it down beforehand and derive the equations of motion. I would dispute that claim, but it was carried further by Jacobi's followers, most notably Mach. In his *History of Mechanics*, already quoted, he stated that the least action principle is basically empty, thereby echoing the objection of Clerselier to Fermat two centuries before: "In every motion, the actual trajectory will always appear as distinctly apart from the infinity of possible ones. But, analytically, it means nothing else than the following: it is always possible to find formulas the variation of which, equated to zero, yield the equations of motion, for the variation can only vanish when the integral takes on a value which is uniquely determined." This is even less clear than Jacobi, but the import is clear: the least action principle tells us nothing, except that the universe is deterministic, that is, that the motion is uniquely determined by its initial conditions.

It is very strange that Jacobi and Mach, always so ready to pounce upon the mistakes of others, should be so definite while making so little effort to explain their arguments or support them mathematically. For in fact, they are wrong: the least action principle, even if one discards the idea of minimizing and reverts to stationary paths, tells us something more than the simple fact that physical laws are deterministic. One could write down deterministic laws of motion which do not follow from a minimum principle, or a stationary path principle. In fact, it is a very interesting mathematical problem, known as the inverse problem in the calculus of variations, to find *all* deterministic laws which follow from a stationary path principle, and to this day it is not fully solved. That the laws holding sway around us are of that type is a challenging fact, which tells us something about the structure of the universe. We shall explore this structure in later chapters, and identify it with a certain geometry of the phase space.

But we will remain on purely mathematical ground. Maupertuis' dream has died with him, and now we shy away from any kind of philosophical interpretation of scientific theories. It may be for want of culture: scientists today are narrow specialists of one tiny field of knowledge, and often have very little experience of life outside universities or laboratories. For all his faults, Maupertuis was a much broader personality; his scientific work extended from biology to mathematics, he was a friend of philosophers and kings, and his experience extended far beyond academia. It may also be for the sake of prudence: in this past century, we saw many ideologies come and go, although they called themselves scientific and claimed to rely on scientific methods. Science itself has undergone many revolutions, quantum physics and molecular biology being the two paramount examples, which have made us acutely aware of the transitory character of knowledge. Galileo, Descartes, and even Maupertuis, who were the first ones aboard, may have had the feeling of discovering eternal truths, but this is no longer the case for scientists today.

We are much closer to Fermat, with his lawyer's training, who refused to commit himself to the meaning of anything, including his quickest path principle. However, he pointed out that, to the best of his knowledge, it worked, and that was good enough. All one can ask of a mathematical model is that it accounts for all the facts that it is supposed to describe, and that it is thrifty in basic assumptions. As Mach put it, science has to "explain facts as accurately as possible, at the least possible intellectual expense"; the idea that there is some ultimate truth to be reached, that all facts can be accounted for with one basic principle, or a few of them, he dismisses as "theological, animistic, or mystical conceptions." In his great *History of Mechanics*, he shows how difficult it was to get rid of these survivals from a time when science did not exist, when humans had to face alone the hardships of nature, and he denounces remnants of these "primitive" conceptions even in the greatest scientists of the eighteenth century:

> When we see that the French Enlightenment philosophers believed themselves to be very close to their goal, which was to account for all of nature with physics and mechanics, and that Laplace imagined some demon who would be able to predict the state of the universe at any time in the future if he were given, at some initial time, all the masses with all their positions and velocities, not only does this enthusiastic overestimation of the physical and mechanical ideas acquired in the eighteenth century seem very excusable to us, but it also is a comforting sight, noble and elevated, and we can sympathize from the bottom of our heart with this intellectual joy, unique in history. Now that one century has gone by, and that we have had the time of

reflection, this view of the world seems to us no more than a mechanical mythology, in contrast with the animistic mythologies of ancient religions.

In Mach's view, there may be no end to scientific progress because there is no goal to be reached. It is a road without an end, but it is worth traveling: "Science does not claim to be a *complete* explanation of the world, but understands it is working toward a *future* conception of the universe." The vision may never be complete, the grand unification of knowledge may stay forever in the future, and we may have to be content with partial views, ever more detailed and ever more disconnected. This is not satisfactory from a philosophical point of view, but it is not a real question for the practicing scientist; that there may be no ultimate truth or meaning is no impediment for day-to-day progress in science.

Henri Poincaré, perhaps the greatest mathematician of the twentieth century, went the last step by stating that science is not about truth, but about convenience: science is just a compact way of stating the facts. Instead of recording the motions of the planet Mars in the starry sky, is it not more efficient to know that Mars and the Earth both move around the Sun on elliptical orbits, and to derive the apparent motion of Mars as seen from the Earth from purely geometric arguments? And is it not even more efficient to apply Newton's law of gravity, so that all the orbits with all their perturbations can be computed from masses, positions, and velocities as observed today? Certainly a similar result could be achieved by using other models, for instance by assuming that the planets and the Sun revolve around the Earth in complicated orbits, but such models would have to be much more complicated to account for the same facts. In a famous book, *Science and Hypothesis*, published in 1902, Poincaré went so far as to state that "these two sentences: 'the Earth orbits around the Sun' and 'it is more convenient to assume that the Earth orbits' have precisely the same meaning." Catholic circles seized this statement as a posthumous vindication of the condemnation of Galileo, and Poincaré had to clarify his thought. In his next book *The Value of Science* (1905), there is a chapter about "science and reality," where he states, "It will be said that science is but a classification, and that a classification can only be convenient, not true. But if it is truly convenient, it will be so not only for me, but for all human beings; it will truly remain so for our descendants; and it is also true that this cannot be due to chance. To sum up, the only objective reality is the relations between things, whence this universal harmony. To be sure, this harmony could never be conceived outside some mind that conceives or feels[these relations]. But nevertheless they remain objective because they are, will be, or will remain, common to all thinking beings."

Twenty years later, Ludwig Wittgenstein will sum it up in lapidary fashion: "The world is the totality of facts, not of things."[21] We observe facts, and we do not know what lies behind them. What a change since the great precursors, Galileo, Descartes, or Newton! They saw the world as a well-designed machine, and they were looking for the blueprints. Mach and Poincaré, on the other hand, see scientists as individuals trying to gather information and store it in the most compact fashion, without much regard to the question whether the consensus they reach has some deeper meaning. Poincaré will not look beyond the layer of facts: "The exterior objects, for instance, for which the word 'object' has been coined, are indeed *objects*, and not fleeting appearances, because they are not only groups of sensations, but groups of sensations related by a constant link. It is this link, and this link only, that constitutes the object behind the sensations, and this link is but a relation." To take an example, a very young child, when shown a bright ball or another interesting object, will lose interest if it is hidden behind a screen; only at a later age will she go or reach behind the screen for the object. In time, she will learn that there is a new ball behind the screen, and since it is always so, she will learn to identify it with the old ball, and simply call it "the ball": an object has been born.

We are a far cry from Plato, for instance, who taught that the objects we observe are but images, or shadows, of originals, the only true and real Objects, which exist in a world above our own, the world of Ideas. After death, the souls of people leave their bodies for that higher world, where they are rewarded or punished according to their deeds, and where they contemplate the true Objects, as well as the great Ideas: the Good, the Beautiful, and the Truth. Then they are sent back to Earth in another body, and they keep a dim memory of what they have seen; the pale and corrupted copies they see in this world make them yearn for something else. Science, then, would be part of the great effort of recovering the lost truth. Poincaré points out that science does not need that kind of belief: there is no need for objects to exist in any other way than to relate our sensations with common experience. Clearly there is no more room for metaphysics: science can be concerned with relating only facts, not things. Quoting Wittgenstein again, "Whereof one cannot speak, one must be silent." Much more would be said about the least action principle, and Poincaré himself was a major contributor. But Maupertuis' dream of science showing a hidden purpose in nature is over.

21. *Tractatus logico-philosophicus* (1921).

From Computations to Geometry

THE COMBINED EFFORTS of Joseph-Louis Lagrange, William Rowan Hamilton, and Carl Gustav Jacobi succeeded in turning the physical insight of Galileo into a coherent and comprehensive mathematical theory. Using the mathematical tools which were developed in the eighteenth century, they discovered a general method to write down the equations of motion for any conceivable mechanical system, submitted to a variety of forces and constraints. Their method is the mathematical foundation of classical mechanics: the equations of motion of any mechanical system (provided there is no dissipation of energy) always are particular cases of the so-called Euler-Lagrange equations.

Neither Lagrange, nor Hamilton, nor Jacobi, nor any of their followers laid much stress on the least action principle. It was a subject best avoided, and if it was mentioned at all, it was described as irrelevant or useless. Lagrange, for instance, stated that the least action principle was "a simple and general consequence of the laws of mechanics," and went on to explain that it was a convenient way of teaching results which have been obtained by other methods. This was the prevailing position until Henri Poincaré, one century later, set classical mechanics on a new course. But the old mistrust survives: even today, I would be hard put to mention a textbook or treatise on classical mechanics which gives more than passing mention to the least action principle.

Lagrange, Hamilton, and Jacobi concentrated their efforts on solving as many problems as possible. That is, for any given problem in mechanics, they wrote down the equations of motion, and then they tried to solve these equations. This is called the "analytical" approach to mechanics, in contrast to earlier methods which relied on geometrical drawings and special properties of curves, as in Newton's *Principles*. In his seminal work, called *Analytical Mechanics*, the first of many treatises by many authors in many years to come to bear that title, Lagrange gave a general method, building upon earlier work by Euler, for writing down the equations of motion, taking into account the internal structure of the moving objects and the external forces and constraints, thereby fulfilling the first part of the program.

The second part was left unfulfilled. Lagrange gave no general method for solving the equations of motion. To solve an equation means to compute the state (position and velocity) of the system at any future time from its state when the motion starts. This can be done either by a computer program (the result is then given as a set of numbers) or by hand calculation (the result is then given in terms of known functions). The former was not available in Lagrange's time, and we are left with the latter: expressing the values of the observable quantities in terms of their values in the beginning and after the elapsed time. Numerous examples of this had been given in the years before Lagrange's time, the most famous one being Newton's solution of the two-body problem, so that Lagrange may well be excused for believing that this was a general feature, and that the equations of motion could always be solved in this fashion, given sufficient ingenuity. In fact, this is not the case, as Poincaré showed a century later: there are only a very few problems in classical mechanics for which the equations of motion can be solved. But this was far beyond the mathematical horizon of Lagrange, who may have hoped that the future would provide a general method for solving the equations which bear his name. Meanwhile, he solved as many problems in mechanics as he could, and left the remaining ones for others to tackle.

Problems for which the equations of motion can be solved are nowadays called *integrable*, in deference to the ancient usage, whereby one talks of "integrating" equations instead of solving them. The first integrable problem is of course the pendulum problem, either in its simple Galilean version, or in the more refined ones studied by Huygens. As we have seen, Huygens, and the other mathematicians working on such problems, such as the Bernoulli brothers or Newton himself, used geometrical methods. Their proofs rely on special properties of certain curves, such as the cycloid or the ellipse, and cannot be extended to other situations. Leonhard Euler was the first to consider the problem in full generality. His book *A Method for Finding Curves Which Are Maximizing or Minimizing*, published in 1744, was read ten years later by the young Lagrange, who devised his own method, for which he coined the name "calculus of variations," and sent it to Euler in a letter dated August 12, 1755 (it should be noted that Lagrange was nineteen at the time). Euler, always a generous man, adopted Lagrange's method and terminology. In the introduction of Euler's 1766 book, *Elements of the Calculus of Variations*, we find an account of theses momentous events:

> All natural approaches to the problem should be free from any geometric consideration. And the greater the hope to open up a new field to calculus, the greater the difficulties to overcome to apply it to this type of problem. Even though I have devoted to this question much time and attention, and I have shared my hopes on this matter with many of my friends, it is this deep mathematician from Torino, Lagrange, who was the first to succeed, and to have reached by pure calculus the same conclusions I had obtained earlier by geometric considerations. In addition, his solution opened up a whole new chapter in calculus, whereby the domain of this discipline was increased substantially.

Euler's book on the calculus of variations is a remarkable work, the first treatise on the subject, and it is quite appropriate that the fundamental equations now are known as the Euler-Lagrange equations. There are two appendixes, which are of even greater interest. The first one studies the equilibrium position of load-bearing beams. It is the first appearance in scientific literature of the phenomenon known as buckling: a vertical beam, on the top of which a heavy load is laid, will stay vertical if the weight it carries is less than a certain critical value. But if the stress crosses the threshold, then the beam suddenly bends sideways (and usually breaks in the process). This phenomenon is much studied nowadays, and it is remarkable that Euler noticed it at such an early stage. In the second appendix, Euler studies the motion of objects in the vacuum, subject to gravity or other forces, and shows how Maupertuis' least action principle, combined with the newly discovered rules of the calculus of variations, yields the equations of motion. It is the first time that the least action principle was used in full generality, Maupertuis himself having never applied it except to very simple cases.

"The most important part of this book," says Jacobi in a 1837 talk, "is a small appendix, where it is shown that for certain problems in mechanics, the trajectory followed by the moving body achieves a minimum (only plane motions are considered here). It is this appendix which gave birth to the whole of analytical mechanics. Some time after its publication, Lagrange, perhaps the greatest mathematical genius since Archimedes, came forward with his *Analytical Mechanics*. . . . By generalizing Euler's method, he discovered his remarkable formulas, which contain in a few lines the solution of all the problems of classical mechanics."

Of course, Jacobi was overly optimistic, for Lagrange's *Analytical Mechanics* contains the equations, not the solutions, as we pointed out earlier. It was nevertheless a tremendous intellectual achievement. Let us recall one more time Lagrange's proud introductory words: "No pictures are to be found in this book. The methods I explain need neither constructions, nor arguments from geometry or mechanics, but only algebraic operations, carried out in an orderly and uniform fashion. All who like calculus will enjoy seeing mechanics becoming one more branch of that discipline, and will be grateful to me for having extended its domain." After Euler's and Lagrange's work, no more pictures were needed, nor any knowledge of geometry; all problems of mechanics could be formulated and the equations written down in a systematic way. *Analytical Mechanics* proceeded to give examples, the most prominent one being the mechanics of rigid bodies.

A rigid body cannot be assimilated to a point, because it has a certain shape. Its position is not given by its location only; one also has to tell how it is oriented in space: is it upside down, or turned sideways? Six numbers are enough to fully define the position of a rigid body, three for the location and three for the orientation. It follows that six more numbers would be needed to specify the velocity: three to tell us how it is traveling through space, and three more to tell us how its orientation is changing. In contrast, three numbers only are needed to give the position of a point mass, and three more for its velocity, six numbers in all, against twelve for a rigid body. Describing the motion of a point mass was the basic achievement of Galileo and his followers, and Lagrange's book gives a complete account. Next in the line of complexity is the motion of a rigid body, and it is certainly much more important, for point masses are mathematical fictions only (imagine a body which is infinitely small and yet carries mass). No wonder attention was turned to studying the motion of rigid bodies very early on.

Unfortunately, that problem cannot be solved in full generality. We now know that one can solve the equations of motion for a rigid body only in very special cases; these are called the *integrable* cases. These cases were discovered through the efforts of generations of mathematicians, and the search for more is going on even today. It may be worthwhile to give a brief account of this exploration.

In his 1760 book *Theory of Motion of Solid or Rigid Bodies*, Euler showed that the motion of a rigid body can be understood as the sum of two independent motions: the center of mass moves as if it were a single point concentrating all the mass of the rigid body, and the orientation along the trajectory is the same as if the rigid body were moving freely around its center of mass. So the general problem splits into two sub-problems: find the motion of a point mass subject to certain forces, and find the motion of a rigid body fixed at its center of mass and subject to certain forces. The first problem is well understood, and we are left with the second one.

Euler solved it in the special case when the forces applied to the rigid body are nil—the case of so-called free motion. In that case, the equations of motion can be solved. Euler's solution will tell us, for instance, how a rigid body will move in intergalactic space: it will travel in a straight line, since there are no forces acting on it, while rolling over itself in a way described by Euler's equations. It will not tell us how a rigid body would fall to Earth, for instance, for in that case there are forces acting, namely gravity. This problem cannot be solved in full generality, and solutions have been found only in special cases where the rigid body satisfies additional symmetry requirements. The first such case was treated by Lagrange himself: it is the case when there is an axis of symmetry (the center of mass then lies on that axis). Tops are usually built according to Lagrange's requirement, and this is why textbooks on classical mechanics still feature detailed studies of spinning tops: not because of a special interest in outdated children's games, but because it is one of the very few cases when the equations of motion can actually be solved. A century later, in 1888, Sofia Kovalevska found another (very special) case of complete integrability, and this is basically it. We cannot solve the equations of motion for rigid bodies under gravity, except for the integrable cases described by Lagrange and Kovalevska.

Let me try to describe in more detail what I mean by this last statement. A solution of the equations of motion for a rigid body (the so-called Euler-Lagrange equations in a particular case) would be a set of twelve mathematical relations giving the position and velocity of the rigid body at any time in terms of the position and velocity when the motion was initiated (or when observations started) and elapsed time. Given any initial position and velocity, these relations define subsequent positions and velocities as functions of time, thereby defining the associated trajectory of the motion. Such relations should, in addition, be computable: it is not enough that they exist; there should also be a practical way (an algorithm) of computing the position and velocity from the initial data for any subsequent time and with any degree of accuracy. This will be the case, for instance, if these relations can be expressed in terms of standard functions, like $y = x^2$ or $y = \sin x$, and this is what Euler, Lagrange, and Kovalevskaya managed to do for some special cases, corresponding to integrable systems.

Unfortunately, as we stated before, these cases are really special. In general, the motion of a rigid body is a nonintegrable problem. This means that we cannot follow all the trajectories of the motion forever, as we could in the integrable case. As the saying goes, one can fool all of the people some of the time, and some of the people all of the time, but one cannot fool all of the people all of the time. Similarly, in a nonintegrable problem, one can follow all of the trajectories for some of the time and some of the trajectories all of the time, but one cannot follow all the of trajectories for all the time. For instance, nowadays we have ways of finding periodic trajectories in nonintegrable systems: such trajectories close upon themselves, which means that the underlying mechanical system indefinitely goes through identical positions and velocities at regular intervals. Periodic trajectories certainly can be followed all of the time, for they are simply repeating themselves, but even the neighboring trajectories, starting from nearby positions and velocities, may very quickly separate from the periodic one and run out of reach of our computations.

In classical mechanics, nonintegrable problems are the rule and integrable ones the exception. They are a very atypical and restricted class. This was not properly appreciated until the work of Henri Poincaré, in the late years of the nineteenth century. All his predecessors concentrated on finding integrable problems, or on investigating mechanical systems which are close to being integrable. This probably was the best that could be done at the time, long before computers were invented, but the drawback of this line of research was that too much familiarity with integrable systems led to the idea that they were somehow typical, and that their properties gave some indication of what happened in more general situations. For instance, it was generally believed that physical systems obeying Newton's laws would have predictable motions and exhibit long-range stability. Today it is well-known that such systems are much more likely to have chaotic behavior. It is a testimony to the power of education that classical mechanics could operate for so long under a mistaken conception. Teaching and research concentrated on integrable systems, each feeding the other, until in the end we had no longer the tools nor the interest for studying nonintegrable systems.

Wrong as it was, this view of mechanics left a deep imprint in philosophical thought, so it is worthwhile to describe it in more detail. What are the main characteristics of integrable systems, and how do they extend to natural sciences at large? First of all, the equations of motion can be solved, meaning that all the trajectories can be computed to any future time and any desired accuracy. As a consequence, any future state of the system can be fully predicted from current data. Not only are integrable systems predictable, they are also stable, which means that any small change in the state (position and velocity) of the system at any given time will lead to similarly small changes at all subsequent times. In other words, for such systems, effects are proportionate to causes: small perturbations, a butterfly flapping its wings for instance, will not develop into major turbulences, such as a tropical thunderstorm.

As we shall see in the next chapter, both these properties, predictability and stability, are special to integrable systems, and we will also see examples of mechanical systems which are unpredictable and unstable. However, since classical mechanics has dealt exclusively with integrable systems for so many years, we have been left with wrong ideas about causality. The mathematical truth, coming from nonintegrable systems, is that everything is the cause of everything else: to predict what will happen tomorrow, we must take into account everything that is happening today. Except in very special cases, there is no clear-cut "causality chain," relating successive events, where each one is the (only) cause of the next in line. Integrable systems are such special cases, and they have led to a view of the world as a juxtaposition of causal chains, running parallel to each other with little or no interference: here I am walking down the street, minding my own business, blissfully unaware that the wind is blowing over the rooftops. Why should I care? The wind belongs to another causal chain, developing independently from mine, according to different rules, and there are many others going on at the same time that I do not have to keep track of. Moreover, I expect that the world is predictable and stable: I will certainly reach the appointment I started out for, and if I am five minutes late now, I will arrive about five minutes late.

But this view may be shattered by an unexpected event: the wind blows off a tile from the roof, and it falls on my head, canceling all future appointments. Two seemingly independent causal chains have turned out not to be independent after all, and this sad event is the result; it might be said to have two causes instead of one, my rushing to an appointment and a particular gust of wind. In the classical analysis which prevailed in nineteenth-century philosophy, this is the place left for chance in a world which is otherwise fully predictable and stable: two independent causal chains may cross each other, and at the intersection we find events which could be predicted neither from one chain alone nor from the other, and which therefore are attributed to chance.

I have never shared this view of the world. In the preceding, classical, example, if one really wants to analyze it in terms of interfering causal chains, one will find not two, but many, perhaps infinitely many. For if that particular tile (not the one right beside it) fell at that particular moment (not before or after), there must have been a reason, which is part of yet another causal chain: it may have been poorly built, or poorly fixed, or someone might have displaced it by walking on the roof, each of these opening up new questions, so that that single event will be shown to be part of an intricately woven tapestry where the threads are causal chains. Why did I have that appointment in the first place? How were the time and place decided upon? Why did the bus driver wait for me as he saw me running to catch the bus? Why did that person stop me along the way to ask me for directions? Each one of these events, if it had turned out otherwise, would have resulted in the tile falling earlier or later, near me and not on me, and each one therefore qualifies for being a "cause" of my death. In fact, almost anything one can think of, provided it happened before the fact, can be connected to the accident by some causal chain, and I could build a case against the whole world for conspiring against me.

The world does not separate into causal chains, arranging events linearly, each one being the cause of its successor and the consequence of its predecessor. Each event is like the trunk of a tree, plunging a network of roots deep into the past, and raising a crown of branches high into the future. There is never a single cause for any event: the deeper one delves into the past, the more antecedents one finds for any particular occurrence. Nor is there a single thread of consequences: the farther one looks into the future, the wider each singular event casts its net. Blaise Pascal once remarked that if the nose of Cleopatra had been shorter, the face of the world would have been changed. Indeed, in a famous episode from Roman history, during the battle for power after the assassination of Julius Caesar, the main pretender, Mark Anthony, was so much in love with Cleopatra that he took her with him during a crucial sea fight against Octavius at Actium, in 31 BC. When she took fright and left the battlefield, Mark Anthony followed her with his own galley, leaving the fleet in disarray at the flight of its commander. After defeating his rival, Octavius went on to become Augustus, the first Roman emperor. If Cleopatra's nose had been shorter, goes the argument, in a time when no aesthetic surgery was available, Mark Anthony might not have fallen in love, and, being much the better commander, he would have won the battle at Actium and succeeded Julius Caesar. What the long-range consequences would have been is open to debate, but one should note that it was Octavius who changed the institutions of the Roman Republic, turning it into an empire which did not disappear from the European scene until 1806.[1]

So long-range history does not split neatly into well-defined causal chains, running parallel to each other; every small incident may turn out to have unforeseen consequences. This is in stark contrast to the mathematical theory of integrable systems. It can be shown (indeed, it is the major result of the theory) that such systems do split into independent subsystems, which never interfere with each other. These subsystems are extremely simple; in fact each of them behaves like a Galilean pendulum. So an integrable system is simply a collection of pendulums, oscillating independently of each other. In this case, the concept of the causal chain is perfectly adequate. Each pendulum represents a causal chain. If I perturb the motion of one pendulum now, the other ones will not be affected, but that one will be; any future change in its position or velocity I can rightly see as the effect of the initial perturbation. Conversely, any global change in the full system splits into a series of changes for each subsystem, each of which has its original cause in some earlier perturbation. These causal chains never interfere, meaning that each pendulum is independent of the others; there is no room for chance in an integrable system. In addition, small perturbations today lead to small perturbations in the future, so that if I want a big change in the future I must inflict a big change today. If the world were an integrable system, the size of Cleopatra's nose could not have had such a disproportionate effect.

Reality lies somewhere between integrable systems, with their orderly succession of proportionate cause and effect, and nonintegrable ones, where everything depends on everything else, and nothing is too small to be taken into account. It is mostly a matter of the time horizon: in the long run, the world is a nonintegrable system. In the short run, if one wants to predict the weather tomorrow, for instance, or the position of the Moon one thousand years from now, integrable systems provide an excellent approximation to reality. At that range, predictions are safe, and we will be told with quasi-certainty whether it will rain tomorrow or whether there will be a solar eclipse in the year 2100. In the long run, however, things are different: we are not sure what the weather will be like in a hundred years (witness the debate on global warming), or where the planet Mercury will be in several billion years (there is a possibility that it may have drifted away from the Sun). The difficulty about making

---

1. When Franz II of Hapsburg, Emperor of Austria, renounced the title of Roman emperor.

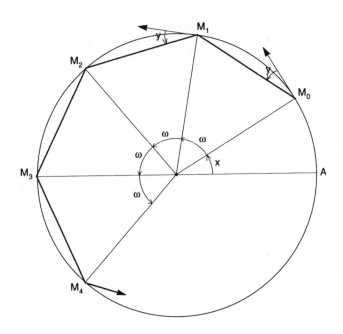

10. THE CIRCULAR BILLIARD TABLE These two pictures represent the circular billiard from two different points of view. The first one shows a single trajectory of a ball starting at M_0. Successive impacts, M_0, M_1, M_2, ... occur at regular intervals ω, so that the nth impact M_n occurs at an angle $n\omega$ from the initial one. Because of the circular shape, at each impact point the ball hits the boundary at

predictions at that range is that there are more and more factors to be taken into account, so many in fact that one does not really know which ones will turn out to be important. This does not mean that long-range predictions are impossible. There is constant progress in understanding underlying physical, chemical, and biological interactions, and in sheer computational techniques, so that the horizon of meaningful predictions is constantly pushed farther into the future. But there will always remain an outer limit beyond which we cannot look, and in many important instances it is still uncomfortably close.

As an illustration of this transition from integrable systems to non-integrable ones, and the progressive breakdown of linear causal chains, we will devote the remainder of this chapter to the simplest possible system in classical mechanics: the motion of a ball on a billiard table. We shall assume that the ball is perfectly round, the bounces perfectly elastic, and we shall neglect the friction of the air and of the felt on the table. Once started, the ball moves indefinitely with constant speed,

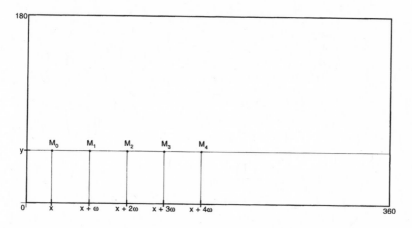

the same angle y, and bounces back at the same angle y, but on the other side of the ray through M_n.

Each impact is characterized by its position and angle. The latter is constant (equal to y), and the position of the nth impact is given by $x + n\omega$, where x is the position of the initial impact. By plotting the successive points $(x + n\omega, y)$ in a rectangle, the horizontal side of which runs from 0 to 360 degrees, and the vertical one from 0 to 180, we get the second picture. The successive points lie on a horizontal line, at the height y from the lower side. If the trajectory closes up after N impacts, that is, if M_N coincides with M_0, then there are only N points on that line. This is the case of periodic motion. If not, then the successive points dot the whole line. This is the nonperiodic case, which is the most general one.

and its trajectory between bounces is a straight line. The bouncing rule is the standard one: the incoming angle (called the angle of incidence, i) is equal to the outgoing one (the angle of reflection, r).

It is the shape of the table which will determine whether the ball's motion is integrable or not, and, as we will see, it makes a huge difference. Standard billiard tables are rectangular, but that will not be the case here. The ideal billiard tables we consider will have neither straight edges nor corners. The edge (let us henceforth call it the cushion in the billiards tradition) shall be one continuous smooth curve. We shall, however, require that any two points A and B on that edge can be joined by a straight line, that is, that the ball can roll directly from A to B without having to bounce off the cushion first. This property is called convexity; all our billiard tables will be convex.

The simplest example is when this edge is a circle. The ball's motion then is quite easy to follow. Bounces occur at regular intervals, separated by a constant angle, say ω. Measuring angles from the first bounce, one will find the second one at an angle 2ω, the third one at an angle 3ω, and so forth. If ω happens to be a true fraction of 360 (all angles here are given in degrees, and 360° is a full circle), that is, if $\omega = 360\,p/q$ for some integers p and q, then this particular trajectory closes upon itself after q bounces; it will then have gone p times around the table. We call such a trajectory *periodic*. If, on the other hand, ω is not a true fraction of 360, that is, if $\omega/360$ is an irrational number, then the trajectory never closes upon itself: the ball goes around and around the table, without ever hitting the cushion twice in the same spot.

One can represent this simple mechanical system geometrically. Consider a rectangle with sides 360 and 90, lying on its long side. Each point of this rectangle corresponds to a pair (x,y), where the first number x gives the position over the long side (it is called the *horizontal coordinate*), and the second one, y, gives the position along the short side (it is called the *vertical coordinate*). In other words, choosing a point in the rectangle is tantamount to choosing two numbers x and y, with x lying between 0 and 360 and y lying between 0 and 90. Let us now interpret x and y in a different way. Choose a point A on the cushion (the edge of the billiard table), and mark it with a notch or a touch of paint; all angles will be measured from A. A pair (x,y) will then represent a bounce, complete with the location of the bounce and the values of the angles of incidence and reflection. The first number, x, gives us a point M on the cushion, namely the only point M where the angle AOM is equal to x (here O is the center of the circular table). This will be where the ball hits the cushion; for instance, $x = 0$ means that the bounce occurs exactly at A, and so does $x = 360$. The second number, y, gives the incoming angle: $y = 0$ means that the ball is glancing in tangentially to the cushion; $y = 90$ means that the ball is hitting the cushion perpendicularly (and bouncing back on its own track).

A trajectory of the billiard ball is nothing but an infinite succession of bounces, each of which is represented by a point in the 360 × 90 rectangle. The first bounce is represented by a point with coordinates (x_1, y_1), where x_1 gives the position of the impact on the cushion, and y_1 the incoming angle. And so on, the nth bounce being represented by a point with coordinates (x_n, y_n), and n ranging through all the integers. In this way, we have a second geometrical representation of the billiard table. Originally, we saw it as an infinite sequence of line segments folded inside a convex box. Now we see it as an infinite sequence of points in a rectangle. The rectangle is, of course, much easier to draw (the picture is less cluttered), and it lends itself much more easily to analysis. Indeed, the impact points are arranged at regular intervals along the circle, so that the sequence of angles x_n is given by the simple rule $x_{n+1} = x_n + \omega$. At each impact point, the incoming angle y_n is the same; by an elementary geometric argument, we find that this angle is equal to $\omega/2$. So we have $y_n = \omega/2$ for every n. Each point (x_n, y_n) of the trajectory is at the same height $\omega/2$ in the 360×90 rectangle.

If we put all this information together, and if we draw the sequence (x_n, y_n) corresponding to a particular trajectory, we find that all these points are located on the same horizontal segment of the 360×90 rectangle, at the height $y = \omega/2$. The behavior of the x_n will depend on the value of ω. If $\omega/360$ is a fraction, say $\omega/360 = p/q$ with integers p and q, then there will be exactly q possible values for x_n, corresponding to q different points on the horizontal segment $y = \omega/2$. The trajectory will run through these q points, and then run through them again, in the same order: it is periodic. If $\omega/360$ is irrational (meaning that it is not a true fraction), then x_n will be evenly distributed on the horizontal segment $y = \omega/2$: if you use a computer, you will see that segment becoming progressively darker as the swarm of impact points settles over it.

Conversely, every horizontal segment in the 360×90 rectangle hosts a whole family of trajectories. They all have the same y, which is the height of the segment, say $\omega/2$. If we go back to the other geometrical representation, the circular billiard table, this means that they all impact the boundary circle at the same angle ω. There is a whole family of them because they differ by the positions of the impacts: given the first one x_1, all the others follow by repeating the relation $x_{n+1} = x_n + \omega$. Going now to the second geometrical representation, we find that all the points (x_n, y_n) fall on the given horizontal segment of height $\omega/2$.

If we now look at this system through the eyes of a philosopher, we see two distinct causal chains operating side by side, without interfering. From the initial impact (x_1, y_1), all the others follow, so that this initial impact can rightly be called the cause of all the other ones. We can be even more precise in our analysis of causality. If we change only x_1, other things (namely y_1) being equal, then only the x_n change, while the y_n stay unchanged. If we change only y_1, other things (namely x_1) being equal, then only the y_n change, while the x_n stay unchanged. In other words, horizontal motion (the x_n) and vertical motion (the y_n) do not interfere: we have two independent causal chains. If we know x_1, then we can predict all the following x_n; the initial value of the other variable, y_1, is irrelevant for that purpose. If we know y_1, then we can predict all the following y_n; the initial value of the other variable, x_1, is irrelevant for that purpose. There is no loss of information during the motion. If we make a small error, of magnitude h, say, on the value of y_1, then we will make the same

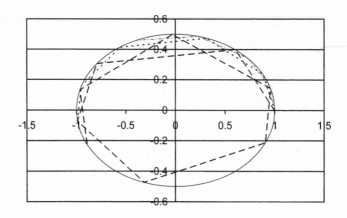

11. THE ELLIPTIC BILLIARD TABLE Its great axis (the horizontal diameter) has a length of 2 units, and its small axis (the vertical diameter), 1 unit. Its two foci F_1 and F_1 have not been represented; they are located on the horizontal axis, on both sides of the center o, at a distance of 0.866. The figure also shows the first impact points of three different trajectories, all starting at the rightmost point, with different angles. Note that the angle of impact is no longer constant along each trajectory, as it was in the circular case.

error on all subsequent values y_n, neither more nor less. Indeed, they all stay on the same horizontal segment in the 360 × 90 rectangle, and our initial mistake simply means that we have misjudged its height.

This is wonderfully transparent, and it is exactly what most people have in mind when they think about causes and effects. Let us now complicate the system a little bit: give the table the shape of an ellipse instead of a circle. A simple way to construct an ellipse is to tie a string to two points F_1 and F_2 (called the *foci* of the ellipse), and to move a pencil along this string while keeping it taut. If the two foci coincide, $F_1 = F_2$, then the ellipse is just a circle. As F_1 and F_2 move apart, the ellipse becomes more and more elongated; it has two diameters, one (along F_1F_2) longer than the other.

The geometry of the elliptic billiard table is different from the geometry of the circular one. Bounces still occur according to the standard rule that the incoming and outgoing angles should be equal, but it no longer follows that these angles should be constant along a given trajectory: each of the successive bounces has its own angle of incidence. Two exceptions are the diameters of the ellipse: if the ball is started along F_1F_2 for instance, it bounces perpetually back and forth along that line, hitting the cushion perpendicularly every time. Three examples are given in figure 11. Another remarkable set of trajectories consists of the ones which go through the foci. If the ball starts from F_1, it has to go through F_2 after the first bounce, then though F_1 after the second, and so on, oscillating perpetually between the two foci. If a room is built in the shape of an ellipse, anyone standing at F_1 will hear distinctly what is being whispered at F_2, a phenomenon due to the concentration at one of the foci of the sound waves emanating from the other, and familiar to the visitors of science museums.

As we did in the circular case, we can represent each bounce by two numbers x and y, where x gives the position of the impact on the cushion and y the incoming angle. In this way, every trajectory of the ball corresponds to an infinite sequence of points (x_n, y_n) in the 360×90 rectangle. It is no longer the case that these points are arranged on a horizontal line: they create more complicated curves, which slice

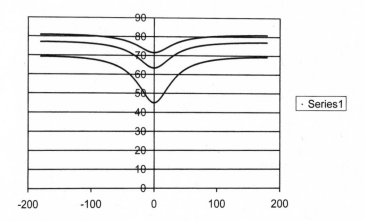

12. THE ELLIPTIC BILLIARD TABLE IN THE POINT/ANGLE REPRESENTATION
In this picture, each impact of the billiard ball on the boundary of the table is represented by its position and angle. The position of the impact is given by its angle from the horizontal axis, as seen from the center of the ellipse, and appears here as the x-coordinate, which ranges from −180 degrees (the rightmost point on the ellipse) to 180 degrees (the leftmost point). The angle of the impact ranges from 0 (normal, or perpendicular impact) to 90 degrees (tangential, or glancing impact). The ellipse is the same one as before, and the three curves correspond to the three trajectories we plotted in the preceding figure. They all start from the rightmost point ($x = 0$) with different angles: 45 degrees (lower curve), 63.4 degrees, and 71.6 degrees (upper curve). The interpretation is straightforward. The first trajectory, for instance, will impact almost all points of the boundary; whenever it comes close to the rightmost point ($x = 0$), its angle will be close to 45 degrees, and whenever it comes close to the leftmost point ($x = 180$ or $x = -180$), its angle with the horizontal will be close to 70 degrees.

through the 360 × 90 rectangle, as the horizontal segments did in the preceding case. Mathematicians call this a *foliation* of the rectangle, and they refer to these slices as *leaves* of the foliation. Three such leaves, corresponding to three different trajectories, are given in figure 12. Leaves are piled upon each other, and each point of the rectangle belongs to one leaf only. The same analysis that we did for the circular billiard table can be transposed to the case of the elliptic one. Each trajectory of the motion stays on one leaf. Therefore, there are really two independent causal chains. The first one determines on which leaf the motion will occur, and the second one determines the motion on that leaf. There is no loss of information: a small error in the initial position just means a small error on the position of the leaf, and this will not change with time.

The elliptic billiard table behaves almost exactly like the circular one. Forecasting, for instance, is easy. Suppose we observe the initial state (x_1, y_1): the position of the impact is given by x_1 and the angle by y_1. We then draw the leaf of the foliation going through the point (x_1, y_1): we know that all subsequent impacts will lie on that curve. This information restricts considerably the future behavior: after all, a curve is a very small portion of the rectangle, and we know that (x_n, y_n) cannot be found elsewhere. To put this in perspective, imagine that we are told that a treasure is hidden somewhere in a rectangle 360 miles long and 90 miles wide. Wouldn't it be much better to know that, in fact, the treasure is hidden along some railroad tracks which happen to cross that region? This is how we will proceed with our forecasting: we first identify the appropriate leaf of the foliation; that is, we find the railroad tracks. We then follow the motion on the tracks.

Alas, all these are very special properties of ellipses (and of circles, which are particular cases of ellipses). As soon as the table has a different shape, the billiard ball behaves very differently. We can still represent any trajectory by an infinite sequence of points in the 360 × 180 rectangle; as before, each bounce corresponds to a pair (x, y), where x gives the position of the impact on the cushion, and y gives the incoming angle. But, in contrast with what happened when the table was circular or elliptic, these points no longer lie on a well-defined curve: they form a cloud. Sometimes the cloud covers the whole rectangle, other times some regions are spared. In the latter case, the cloud does not taper off, there is no smooth transition from a dense cover to a lighter one, and then nothing: the boundary is always sharp. Pictures like figure 13 would call to mind the effect of buckshot hitting a target, or a handful of sand strewn upon the ground, if it were not for these strangely sharp boundaries.

This striking difference means that the system is no longer integrable: in mathematical terms, a nonelliptic billiard table is a nonintegrable system. The difference can be told at a glance, simply by comparing figures 14a and 14d, both of which represent a single trajectory of the billiard ball: there is no mistaking an integrable system with a nonintegrable one. But it goes much deeper than simple appearances: everything we said about causality and prediction for the elliptic billiard table breaks down. Suppose, for instance, that a trajectory starts from a point (x_1, y_1), which represents the first bounce: what can we say about the nth bounce? Well, it lies somewhere in the cloud, and if the cloud covers the whole rectangle, this is no help at all: there are no railroad tracks to save us, as in the case of the elliptic billiard table. If we really want to know something about the nth bounce, for instance its position on the cushion, the best we can do is to painstakingly retrace the whole trajectory: from the first bounce (x_1, y_1) figure out the second (x_2, y_2); from the second one, the third (x_3, y_3); from the third one, the fourth (x_4, y_4); until we reach the particular bounce n we are interested in. Note that the information does not split: even though we are interested only in x_n, the position of the nth impact on the cushion, we need compute not only the positions x of the intermediary bounces, but also their incoming angles y.

It so happens that nonelliptic billiard tables are *chaotic* systems: not only do computational errors accumulate, they blow up. As we compute the second impact, (x_2, y_2), we have to do some rounding off. It is impossible, and not even physically meaningful, to give the values of x_2 and y_2, as the mathematician would want them, with an infinite succession of numbers after the decimal point. There has to be a cutoff somewhere, and this is indeed what computers do. The higher the precision required by the user, the farther away the cutoff, but it is always there. But cutting off the mathematical value means that some error, however small, has been made deliberately. Not only will this error be carried over to the next (third) bounce, and then on to the fourth, the fifth, and so on all the way, but it will also compound at a large rate, perhaps doubling at every step, so that eventually it will dwarf everything else, except, of course, all the other rounding errors which are being made at every step and which grow at the same rate. Very quickly, after as few as ten bounces, the computations are completely drowned in background noise, and the results we get no longer have anything to do with reality. In other words (in contrast, again, with what happens when the table is elliptic), long-term predictions for such billiard tables are impossible.

In another contrast, the system no longer splits into independent causal chains. We saw how, in the elliptic case, the 360×90 rectangle was traversed by curves (we referred to them as railroad tracks), and that if a trajectory started on one of these curves it stayed on it forever. Finding a trajectory then naturally splits into two problems, which can be solved independently: finding the right railroad track, and then finding the motion along that track. There is nothing like that in the nonintegrable case. As we have seen, there is no other way of finding a trajectory than actually computing it, from the first bounce to the one we are interested in, and incurring considerable loss of information at every step. The information does not split: to compute *either* x_2 or y_2, we need to know *both* x_1 and y_1. There is no way of arranging the calculations in two separate channels, each of which would compute one-half the information without

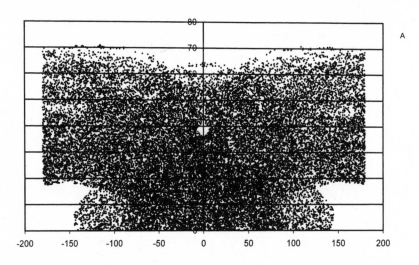

A

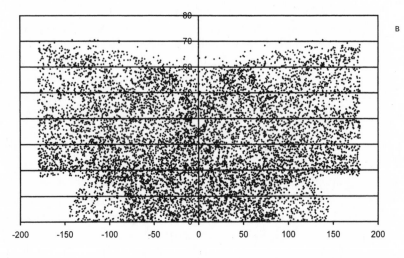

B

using the other half. In philosophical language, there are no longer two independent causal chains: everything matters to everything else. The system can only be understood globally.

There is no greater contrast than between elliptic and nonelliptic billiard tables. On the one hand, a completely transparent system, fully predictable, ruled by two independent causal chains. On the other, an unpredictable system, where no single cause can be found for a single event: everything, always, has to be taken into account. The first is typical of integrable systems, which have been explored by Lagrange, Jacobi, and all the founding fathers of classical mechanics. The second is typical of nonintegrable ones, which were first discovered by Poincaré around

c

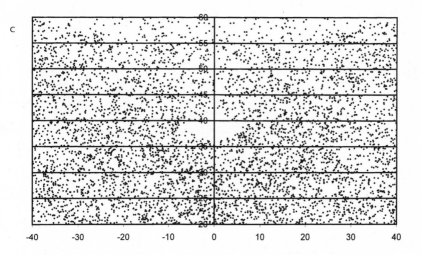

13. A GENERAL BILLIARD TABLE Figure 13a is the point/angle representation of 30,000 impacts of a single trajectory of a billiard ball, when the table is convex, but not elliptic (think of an egg-shaped table). The starting point lies at the rightmost point of the table, and the initial angle is about 56.3 degrees ($x = 0$, $y = 56.3$). It is a good example of a nonintegrable system, and the motion is obviously chaotic. This is very different from the situation of figure 12, when the motion was integrable. It is no longer true that when the trajectory comes back near $x = 0$, the angle will be close to $y = 56.3$; as the figure shows, the trajectory comes back near $x = 0$ a great many times, and the impact angles are all different, ranging from $y = 0$ to $y = 65$. The trajectory explores a large fraction (but not all) of the (x, y)-rectangle. To give a sense of the speed of propagation, figure 13b shows the initial 10,000 impacts of the same trajectory. To give a sense of the inner structure of the cloud, figure 13c gives a detailed picture of the region $-40 < x < 40$, $20 < y < 60$.

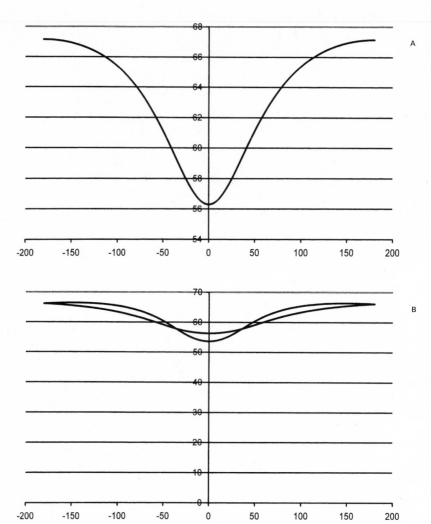

14. FOUR STEPS TO CHAOS These four pictures describe 30,000 impacts of a single trajectory, always starting at the rightmost point of the table with an angle of 56.3 degrees, as the table changes from a purely elliptic shape to an ovoid shape. Figure (a) is the elliptic case, and it is integrable, as we have already observed. In figure (b), the table is no longer elliptic, but is still close enough

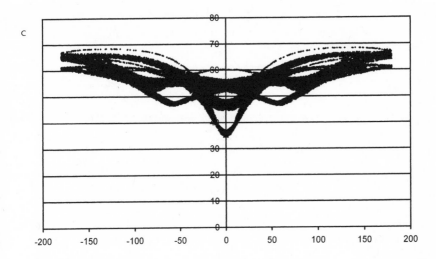

C

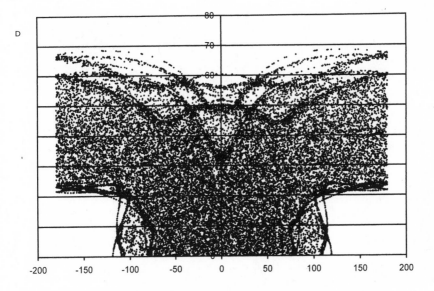

D

for the system to remain integrable; note that at every point of the boundary, there are now two possible impact angles, instead of one in the elliptic case. In figure (c), the table is further away from being elliptic, and chaos sets in. Chaos is fully developed in figure (d).

1900, and which would have remained uncharted territory if the invention of computers had not given us the means to explore them. Which are closest to the real world?

There is no doubt about the answer, for integrable systems are brittle: they break as soon as one touches them. If the billiard table deviates ever so much from being elliptic, the system is no longer integrable, and chaos sets in. The same thing happens if the table is not perfectly horizontal; the slightest bump, the most modest tilt are enough. The transition from integrable to nonintegrable systems is quite interesting to observe. See, for instance, figure 14, which depict four steps toward chaos. Each step depicts a single trajectory of the billiard ball, starting from the same initial bounce, but on four different tables. We start from a perfectly elliptic one, and we deform it progressively. As we drift away from the elliptic case, we see the trajectory fattening up: in the first picture it is a neatly drawn curve, in the last one it has become a cloud, gathered around the initial curve, but covering a large section of the 360 × 90 rectangle. This means that the system has become nonintegrable, and chaotic, but there still remains some predictability: after all, there are some regions of the rectangle that the trajectory cannot enter. In the second picture, for instance, where the perturbation is quite small and the table is very close to being elliptic, we see that, even though the trajectory is allowed to leave the "railroad tracks" of the first picture, it cannot wander very far from them. It has to stay within a certain distance of the tracks, and the smaller the perturbation, the smaller the distance; if the perturbation is zero, that is, the table is elliptic again, then the trajectory is back on the tracks. If we travel in the opposite direction, and increase the perturbation, making the table less and less elliptic, the allowed band along the railway track will increase until it finally covers the whole 360 × 90 rectangle. Seeing the allowed region growing from a neat curve to the whole rectangle highlights a whole range of nonintegrable systems, drifting progressively into unpredictability and chaos.

This is precisely when the least action principle—better referred to, after the work of Hamilton and Jacobi, as the stationary action principle, or, in honor of its discoverer, Maupertuis' principle—comes into its own. As Lagrange pointed out, it is little better than useless if we are concerned with integrable systems, for in that case we can compute all trajectories to any desired accuracy up to any length of time, way up to infinity. Anything we would ever want to know shows up in the calculations; there is nothing left for the stationary action principle to tell us. At best, it enables us to explain known results in a different way. But when the systems under investigation are not integrable, computations quickly break down, and there is very little they can tell us about trajectories. As we shall see, Maupertuis' principle then enables us to pinpoint some of these trajectories and to follow them with mathematical accuracy.

Nonintegrable systems, nonelliptic billiard tables for instance, do not have general solutions; this means that there is no direct way to compute their state at time t from their state at time 0. There remains always the possibility to compute their state at time t by actually following the trajectory of the motion between 0 and t, but this procedure is fraught with rounding errors which accumulate over time, and which fairly often are amplified by the system. For integrable systems, there are shortcuts which give the result directly, and which avoid these pitfalls. But in the nonintegrable case, there is no shortcut, and one has to compute each trajectory individually; as a consequence, our range of investigation is quite limited. At the end of the nineteenth century, the path opened by Lagrange, Hamilton, and Jacobi had been fully explored, and yet the long-term dynamics of most mechanical or physical systems remained shrouded in mystery.

(CHAPTER 5) Poincaré and Beyond

THE GREAT FRENCH MATHEMATICIAN Henri Poincaré (1854–1912) was the first one to investigate successfully the long-term dynamics of planetary systems. This is known in mathematics as the three-body problem: two planets, one large and one small, gravitating around a star. The star is supposed to be large enough to be insensitive to the attraction of the two planets, and the large planet is supposed to be large enough to be insensitive to the attraction of the small one. The large planet then moves around the star in an elliptical orbit, according to the three laws of Kepler, and the whole problem then consists in determining the motion of the small planet. It was a very old problem, and a very important one, since it was the first step away from the two-body problem, which was completely solved by Newton. In addition, solving this problem would help astronomers understand the motion of the Moon: neglecting the attraction of the other planets on the Sun/Earth/Moon system, and the attraction of the Moon on the Earth, yields a three-body problem.

In 1887, Oscar II, King of Sweden and Norway, under the influence of the Swedish mathematician Gösta Mittag-Leffler, created a one-time mathematics prize to celebrate his sixtieth birthday. This competition attracted much attention in the scientific world, and Poincaré won it by submitting a memoir on the three-body problem. In this memoir, he essentially proved that the three-body problem was amenable to the old methods of Lagrange, Hamilton, and Jacobi, and that the long-range dynamics were quite regular. After the prize had been awarded, a young mathematician named Edvard Phragmen was assigned to proofread the prize-winning memoir and prepare it for publication. He found a mistake in the proof of the main theorem, and went to Mittag-Leffler for clarification. Mittag-Leffler wrote to Poincaré, and after a while, Poincaré's answer came back: not only was the proof wrong, the result itself was wrong! Of course, Poincaré was devastated, and worked night and day to completely rewrite his memoir, getting opposite results from the one he had found before. But by this time, the first memoir had gone to press. Mittag-Leffler recalled all the copies that had been sent around the world, and the memoir was reprinted in its new version at Poincaré's expense and resent to the subscribers of *Acta Mathematica* in 1890.

To circumvent the difficulties he had missed the first time around, Poincaré had to introduce completely new mathematical methods. Later on, he developed them in his famous three-volume book, *The New Methods of Celestial Mechanics*, published between 1892 and 1899, which remains to this day a classic of mathematics, especially the last volume, which lays down the foundations of modern chaos theory.

One of the new ideas Poincaré introduced was to use the stationary action principle to find closed trajectories of nonintegrable systems. These trajectories are the ones which close back upon themselves, meaning that if they have gone through a certain state, they will come back to it at regular intervals, like the hands of watch. The common length of these intervals is called the *period*, and the corresponding motion is called *periodic*. The trajectory of the Earth around the Sun, for instance, is approximately periodic, with a period of one year. If there were no other object than the Earth to gravitate around the Sun, then this trajectory would be exactly periodic, and all Kepler's laws would hold. But the presence of the Moon, and of other planets as well, distorts the beautiful Keplerian motion in various ways, so that the resulting motion fails to be periodic.

The New Methods of Celestial Mechanics is entirely devoted to the so-called three-body problem, which consists in describing the possible motions of three objects (say a sun and two planets, or a sun, a single planet, and its moon) attracting each other according to Newton's law of gravity. Although three objects only are involved, it is already an extremely complicated problem, and there are very many possible motions, most of which are not periodic. Why, then, go to all that trouble to find periodic motions? Why would such rare, and nontypical, trajectories be interesting? In the opening pages of his book, Poincaré asks the question with characteristic frankness, and gives a poetic and oft-quoted answer: "The reason the periodic solutions are so precious to us is that they are the only opening by which we can enter this hitherto inaccessible fortress."

The fortress he is alluding to is the long-term behavior of nonintegrable systems. Ever since Newton, mathematicians and astronomers had been trying in vain to have a theory of the Moon, for instance, which would enable them to predict eclipses much farther ahead than the few measly centuries they could reach with their calculations. But the attraction of the Sun on the Moon turned this into a three-body problem, and one ran quickly into all the difficulties of making long-range prediction in nonintegrable systems. What Poincaré is pointing out, in this famous quotation, is that periodic motions are the only ones which can be figured out completely, without fearing that the computational errors would accumulate or amplify, drowning the exact solution in background noise. Indeed, since these motions are periodic, they go through the same states at regular intervals, as a swinging pendulum comes back to the same position with the same speed. A periodic motion will come back exactly to its initial state, once (after one period has elapsed), twice (after two periods), three times, in fact, indefinitely: it never stops. So, if we have been able to figure out the initial state with an accuracy of 1/1000, say, there is no question that, ten billion periods later, the system will come back to the very same point, which is still known with an accuracy of 1/1000: the error does not grow with time.

A beautiful argument, no doubt. But how is one to know a priori that the motion is periodic? Are we not running into the same computational problems which have been plaguing us for so long? If for instance we merely follow a certain trajectory, and the computer tells us that it closes upon itself after a certain time, this is not good enough to conclude that the corresponding motion is periodic. For the computer makes rounding-off errors, and all it can actually tell us is that the initial and final states coincide up to the first sixty (say) decimals. It can tell us nothing about the remaining decimals, which may very well be different, and that slight difference may increase with time, so that after going round a hundred times, say, the trajectory drifts away from its initial state, and is seen, in fact, not to close upon itself.

The stationary action principle pinpoints periodic motions geometrically, without recourse to computations. By doing so, it avoids the pitfalls we have described, and has now become one of the most useful tools in the investigation of nonintegrable systems. We will show how it works by applying it to a simple example. Let us go back to the billiard table and find out its closed trajectories by a general geometric method, applicable to any kind of table, elliptic or not.

We will start with the simplest periodic motions of the billiard ball, those where it bounces back and forth between two points. An example is provided by the segment AB in figure 15: the ball hits the cushion at A, bounces back along AB, hits the cushion again at B, bounces back along BA, and so forth. For this motion to be true to the laws of reflection, the incoming angle must be equal to the outgoing one, which implies that the segment *AB* must be perpendicular to the cushion at its extremities *A* and *B*. Any segment with this property is called a *diameter* of the billiard table. If the table is circular, any segment going through its center is a diameter, and all diameters are equal. If the table is elliptic, there are exactly two diameters, one larger than the other. We will prove that any convex table still has two diameters, and so that the corresponding billiard ball has two periodic motions, with exactly two bounces per period.

The large diameter is easy to find. Put on the billiard table two points M_1 and M_2, and try to push them as far away from each other as possible. There is exactly one position where the distance between them is greatest (exchanging M_1 and M_2 gives another one, but we do not count it as different). Denote this position by *AB*: this is our large diameter. It is intuitively clear, and it can be proved mathematically, that *AB* is perpendicular to the edge of the table at its extremities.

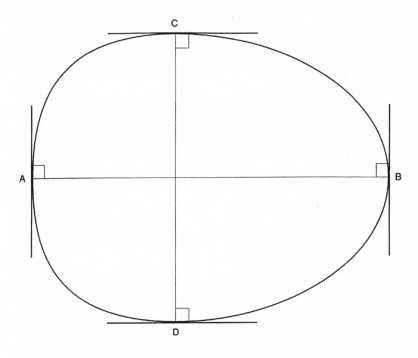

15. DIAMETERS A table is convex if from every point on the boundary one can go to every other point in a straight line without hitting the boundary in between. Such a table always has two diameters, a great one (*AB*) and a small one (*CD*). The points *A* and *B* are the two points on the boundary which are the farthest apart. Is there a similar characterization for *C* and *D*?

Finding the small diameter is quite another matter. Looking at figure 15, it is intuitively clear that there must be another position of $M_1 M_2$ where it is perpendicular to the edge of the table. But how can we find it? The first idea which comes to mind is to minimize the distance between M_1 and M_2 instead of maximizing it, in other words, to find a position where M_1 and M_2 are as close as possible. Such a position is easy to find, and there are many of them: just put M_1 and M_2 together, $M_1 = M_2$. Then their distance is zero, and this is as small as it gets. The segment $M_1 M_2$ is just a point, instead of the diameter we were looking for. So this argument does not work.

We need something else. What we need, in fact, is a theorem which is quite typical of modern mathematics,[1] and which we will state in every-day language: in an island with two peaks (or more) there must be one mountain pass (or more). This is known as the *mountain pass theorem*. One way to see why it is true (although we are still far from a mathematical proof) is to imagine a path crossing the island between the two peaks. Over the centuries, this path will have evolved to lie as low as possible (no one wants to climb high into the mountain if it can be avoided). The path climbs up the slope, reaches its highest point, and then goes down the other side of the mountain. That highest point has to be a mountain pass.

We can do even better: we can relate very precisely the number of passes with the number of peaks. This may seem strange: an island with three peaks can have either two or three passes, so it would seem that the number of passes is unrelated to the number of peaks. The key, however, is to observe that in the first case the three peaks are aligned, so that water can flow directly down the slope of the mountain into the sea, while in the second one the three peaks surround a region out of which water cannot flow, so that a lake forms. The formula we are looking for relates the number of passes to the number of peaks and the number of lakes, namely:[2]

Number of passes = Number of peaks + Number of lakes − 1.

As an example, there are islands consisting of a single peak jutting out of the sea. Sure enough, these islands have neither passes nor lakes, and the formula then reads 0 = 0. As another example, an island with three peaks and no lakes has two passes, but if it has one lake it must have three passes and if it has two lakes it must have four passes (please try to

1. This theorem goes back to the pioneering work of the Russian mathematicians L. A. Lyusternik and L. Schnirel'man.
2. This formula has a long history, starting with Euler and ending with Morse.

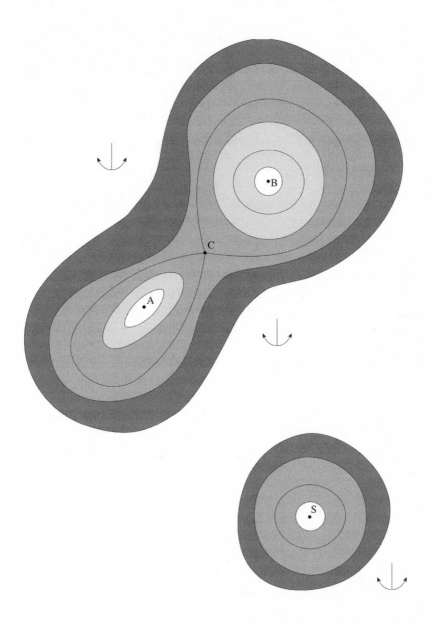

16. ISLANDS On the top island, there are two peaks *A* and *B*. Between them, there must be a mountain pass: this is the point where the level lines cross. On the bottom island, there is a single peak, so there cannot be any mountain pass.

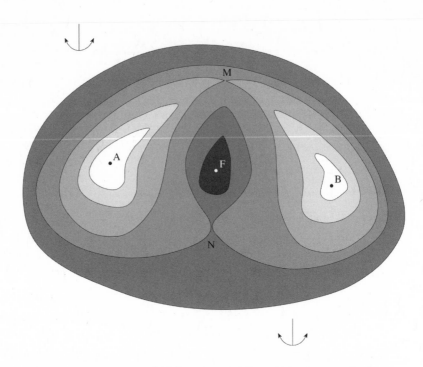

17. THE EULER FORMULA This is a more complicated island: two peaks, *A* and *B*, and an interior basin *F*. The Euler formula calls for $2 + 1 - 1 = 2$ passes, and they are represented at *M* and *N*.

imagine the island; the fourth pass lies between the two lakes). The more lakes there are, the more passes you get.

The reason why such passes are of interest is as follows. They are not of maximum height (the peaks lie higher), nor are they of minimum height (they lie way above sea level), but they share with maxima and minima a special property: at every mountain pass, the ground is horizontal, as it is at the summit of a rounded peak or at the bottom of a rounded pit. Ideally, if a ball is put precisely at the pass, then it will stay there, in equilibrium; if it is shifted ever so slightly, or if a gust of wind pushes it, it will start rolling down one side of the mountain. But at the pass itself, it stays, as if it could not make up its mind in which direction to fall. This is the geographical equivalent of a stationary point:[3] a mathematician would express the above properties simply by saying that a mountain pass is a stationary point of height (whereas a peak is a maximum point of height).

3. See the preceding chapter for a definition of stationary points.

So finding stationary points of the action amounts to finding passes in certain mathematical mountains, and this is why the mountain pass theorem is very useful when one tries to apply Maupertuis' principle. In the case of the convex billiard table, for instance, one can construct a certain island with two peaks, the first one corresponding to the large diameter AB, and the second one to the same diameter, taken in the other direction BA. As we have seen, there must then be a mountain pass somewhere on the island. It corresponds to a position of M_1 and M_2 such that $M_1 M_2$ is perpendicular to the edge of the table, that is, to the small diameter of the table.

The precise argument, for the mathematically inclined, is put forward in appendix 1. The main point here is that it is purely geometrical: there are no computations at all, no algebraic calculations, just an argument about the general shape of islands. It is "soft" geometry, meaning that nothing in the argument relies on some distance being precisely equal to another one, or some angle assuming a certain value, or some line being straight. The classical, "hard," geometry of Euclid, on the other hand, is all about circles and triangles, and its results rely crucially on certain sides or certain angles being equal. Here, making the peaks higher, rounding them or shifting them around will not make the mountain pass disappear. This change from hard to soft geometry is very typical of modern mathematics, and it was heralded by Poincaré in his great treatise on celestial mechanics: the "new methods" he is alluding to in the title are soft, qualitative, geometrical, methods, as opposed to hard, quantitative, computational ones.

So, using the mountain pass theorem, we have found the small diameter we were seeking.[4] It corresponds to a periodic motion, with the ball bouncing back and forth along that diameter. Other periodic motions can be found by the same method. If, for instance, we arrange three points M_1, M_2, and M_3 along the edge so that the perimeter of the triangle $M_1 M_2 M_3$ is the greatest possible, then there is a closed trajectory which bounces three times during every period, once at M_1, then at M_2, and finally at M_3. Again, in each case a closed trajectory will be found by

4. We may have found more than we were looking for. Indeed, the mountain pass theorem tells us that there must be at least one pass on the island, but, as we saw, there may be several. In that case, each of these passes corresponds to a position of M_1 and M_2 such that $M_1 M_2$ is perpendicular to the edge, that is, to a small diameter of the table. Thus there may be several small diameters with different lengths, corresponding to the heights of the different passes. For that matter, there may be more than two peaks on the island; in that case, each of them corresponds to a large diameter of the table. So there may be several large diameters with different lengths, corresponding to the heights of the different peaks. Think for instance of a lozenge rounded off at the corners: it has two large diameters and two small ones.

maximization, but there will also be a second one, which has to be found by a mountain pass argument, and which corresponds to a stationary point of the total length. The final result is that, for every prime number p, and every number q, there are two closed trajectories which, during every period, bounce exactly p times off the cushion and go exactly q times around the billiard table.

Admire, again, the power of this method. It can be applied to any kind of billiard table, provided only that it is convex, that is, that any point on the cushion can be hit directly from any other point. It requires no computation, and there is no fear that any initial mistake we made in locating the initial bounces will accumulate or amplify with time. This is all the more remarkable because, as we mentioned before, nonelliptic billiards are chaotic systems, and the periodic solutions themselves are unstable. If we do not start precisely on the periodic solution, but ever so close away from it, the corresponding motions will drift apart, slowly at first, then faster and faster, and will end up having no relation with each other. Finding such unstable periodic solutions is a real tour de force, which shows all the strength of the stationary action principle.

In the case of the billiards, the action, as defined by Maupertuis, is just the length of the trajectory. What about more complicated systems, with a more complicated definition of the action? Can the stationary action principle still be used to find periodic motions? Imagine for instance that the billiard table no longer is flat, but strewn with bumps which the ball has to climb and pits down which it speeds. Imagine that it is surrounded by a rounded wall, as in skateboard rinks, so that the ball does not bounce off the wall but scales it before falling back on the table. Trajectories will no longer be straight lines bouncing off at sharp angles; they will be smooth curves, turning away from the bumps and into the pits, and swerving back into the field as they reach the boundary, very much like a skateboarder. The speed will no longer be constant; the ball will speed up downhill and slow down uphill, and the faster it goes the higher it will climb on the limiting wall.

Finding periodic motions in such cases is much more difficult than in simple billiards. There are two main difficulties. The first is that one now has to take into account the speed of the ball: it is no longer the case that if you send the ball in the same direction with two different speeds you get the same trajectory. The speed will not affect the way the ball bounces off the cushion, but it certainly will affect the way it comes out of the limiting wall. In the case of simple billiards, every trajectory could be fully described by giving the position of the first bounce on the cushion and the direction of impact: these are two numbers. In the second case, there are no more edges to bounce from, just a rounded wall on which the ball climbs, so giving its initial position now requires two numbers, one to tell in what part of the wall the ball is, and the second to tell how high it has climbed. In addition, to fully specify the motion, we need to give the direction and the speed of the ball, two numbers again. Two numbers for the initial position, and two for the initial velocity—this means that we now need four numbers to specify the initial state, as opposed to two in the case of the simple billiards. The equations of motions will correlate these four numbers. In other words, going from the simple billiards to the complicated one brings us from two dimensions to four. Really complicated mechanical systems can have many more dimensions; four is enough, however, to deprive everyone but the trained mathematician of the help of intuition.

The other main difficulty is that we no longer have an inkling of what closed trajectories look like. In the case of simple billiards, we knew that these trajectories were just line segments connecting the points of impact on the cushion. There is nothing like that in the more complicated case, where the ball no longer moves in a straight line nor bounces off a cushion. This means that trajectories will be much harder to describe: one can no longer be content with finding the impacts on the cushion. All closed curves drawn inside the billiard table are possible candidates for the trajectory, and one has to discover which one or ones satisfy the stationary action principle.

In Poincaré's time, these difficulties could not be overcome. Today, about one century later, the requisite mathematical tools have been developed, so that we now can apply the stationary action principle to very general systems. In the meantime, however, there were unexpected consequences, the most striking of which was the discovery by Michael Gromov, in 1980, of an uncertainty principle in classical mechanics. Heisenberg's uncertainty principle in quantum physics is well known, but no one had ever suspected that there was a similar principle for classical physics. It is still too recent to be known outside a small circle of specialists, but once it diffuses into the scientific community, I am sure it will attract as much attention as its predecessor in quantum physics. Anyway, it is a success story for modern geometry and the stationary action principle, and it is well worth relating.

We will set out the theory using billiards. Take a convex table with a cushion along the edge, a single ball bouncing off the cushion. As we saw before, any trajectory is then completely specified by two numbers x and y, where x gives the position of the impact on the cushion, and y the incoming angle. Starting from the initial bounce (x_1, y_1), we deduce the second one (x_2, y_2), then the third (x_3, y_3), and so forth, so that a single trajectory reduces to an infinite sequence of points in the 360×90 rectangle; this is what we called the second geometric representation earlier in chapter 4.

But now we introduce a new idea. The initial x_1 and y_1 cannot be measured with infinite accuracy: however precise we are in our measurements, there is a limit, depending on the instruments we use, below which we cannot go. So the true initial position x and incoming angle y are not the x_1 and y_1 we recorded, but can be any values in some interval around x_1 and y_1. Denote by Δx_1 and Δy_1 the lengths of these intervals. If we now represent the pair (x, y) by a point in the 360×90 rectangle, then the true value (x, y) lies within a small rectangle, of length Δx_1 and width Δy_1, centered at (x_1, y_1), which we will refer to as the *uncertainty region* around the measured value (x_1, y_1). The smaller the uncertainty region, the more accurate our measurement has been. It is natural to consider the area of this region, that is, the product $\Delta x_1 \Delta y_1$, as a measure of this accuracy; we shall refer to this number as the *uncertainty* on the measured value (x_1, y_1).

No further measurements will be made, and we will merely compute the trajectory of the billiard ball. We will assume that we can perform these computations with infinite accuracy. As we saw earlier, this is not possible in practice, because computers cannot process infinitely many digits, and have to round off decimals at some point. But let us conduct a thought experiment, imagining for instance that the good Lord has lent us for the occasion his own computer which performs all calculations with infinite precision at every step. This being so, the only possible source of error is the one we made in our initial measurement: that one we will have to carry throughout the computation.

To be precise, starting from the measured values x_1 and y_1, we find the position x_2 and the angle y_2 at the next step. Since the actual values of (x, y) for the first bounce are not exactly (x_1, y_1), but are located in the uncertainty region around this point, the actual values of (x, y) for the second bounce will not be exactly (x_2, y_2), but will be located in some uncertainty region around this point. There is no reason why this second uncertainty region should be a rectangle, even if the first one is: its shape will usually be deformed, stretched in one direction and compressed in another. However, it is a remarkable fact, first discovered by the French mathematician Joseph Liouville in the nineteenth century, that the area is unchanged.

Although the shape of the uncertainty region around (x_2, y_2) is no longer a rectangle, we will still denote its area by $\Delta x_2 \Delta y_2$, and refer to it as the uncertainty on (x_2, y_2). Note, however, that the numbers Δx_2 and Δy_2 no longer have any meaning by themselves. Liouville's discovery then can be expressed as the simple equality $\Delta x_1 \Delta y_1 = \Delta x_2 \Delta y_2$. This mathematical relation expresses the fact that the uncertainty is carried over from the first bounce to the second; there is no improving the initial information, nor does it decay (bear in mind that we are still using this divine computer which does not need to round off infinite sequences of digits after the decimal point). Uncertainty will also be carried over to the third bounce, the fourth, and so on: at every step n, the relation $\Delta x_1 \Delta y_1 = \Delta x_n \Delta y_n$ will hold. The uncertainty stays unchanged throughout time, unless, of course, we make a new measurement, using better instruments, thereby diminishing $\Delta x_n \Delta y_n$. Let us express this fact, somewhat loosely, as follows:

First uncertainty principle in classical mechanics: information cannot be created. Only measurement, not computation, can reduce uncertainty.

Let us investigate some consequences of the first uncertainty principle. For instance, it is impossible to design a billiard table that would focus the ball, that is, that would allow us to predict its future position and direction with greater precision than we had at the start. A simple way to see this is to note that the uncertainty $\Delta x_n \Delta y_n$ is pegged at its initial value $\Delta x_1 \Delta y_1 = u$. If we devise a table so that, at some future stage n, the position x_n of the impact can be predicted extremely accurately, so that Δx_n becomes very small relative to u, then, to keep the product $\Delta x_n \Delta y_n$ constant, Δy_n must become very large relative to u, and the prediction on the incoming angle y_n must be very poor.

Unfortunately, this argument, although quite convincing, is incorrect, because there is no reason why the uncertainty region around (x_n, y_n) should look like a rectangle, and therefore it is not clear what meaning to attach to Δx_n and Δy_n. We can salvage it by imagining a pocket around (x_n, y_n) in the 360×90 rectangle, for instance a rectangle with length Δa and width Δb, so that its area is $u = \Delta a \Delta b$, and a player trying to send the ball into it. His initial shot should be (x_1, y_1), which reaches (x_n, y_n), in the middle of the pocket, after n bounces. Unfortunately, the player, although he knows (x_1, y_1), cannot manage that shot exactly, only some approximation of it. The uncertainty $\Delta x_1 \Delta y_1 = s$ around that initial shot measures the player's skill: the smaller the uncertainty, the more skilled the player is.

After n steps, the uncertainty region around the predicted value (x_n, y_n) still has area s, to be compared with the area of the prescribed pocket, u. If s is smaller than u, that is, if the player is not skilled enough, there is no way of fitting the uncertainty region entirely inside the pocket: it is simply too big. So part of that uncertainty region around (x_n, y_n) must remain outside the pocket, meaning that the corresponding shots will miss the target. The player himself cannot tell the difference between these shots and those that reach the target: they all start within his uncertainty region. He hits all shots the same way, but only a certain percentage of them succeed: he cannot consistently hit the target.

This argument does not depend on the shape of the billiard table, and so we are led to our conclusion: there is no designing a table that would focus the ball toward a prescribed target. In other words, the first uncertainty principle tells us that no table will compensate for lack of skill.

Let us now go one step further and imagine not one but several balls moving on the same billiard table. Say there are N of them, moving merrily about. It is not true any more that the trajectories of each ball, between two collisions on the cushion, are linear, and that the speeds are constant: they can collide with another ball in the middle of the table, whereupon each of them proceeds in another direction with another speed. Directions and speeds after the collision are fully determined, just as they are after a bounce on the cushion, so that the entire trajectories of the N balls are fully determined by their initial positions and velocities.

These initial positions and velocities are not known exactly: for each of them, there is a certain region of uncertainty around the measured values. The area of this region is called the initial uncertainty, as before. The initial uncertainty on the nth ball, for instance, will be denoted by u_n, and has the same interpretation as before: the smaller u_n, the more accurate the measurement of the initial position and velocity.

The first uncertainty principle applies not to each u_n individually, but to their sum, $u_1 + u_2 + \cdots + u_N$, which we denote by U, and which we call the *total uncertainty*. More precisely, it is the total uncertainty at the initial time, $t = 0$, when the motion starts; but, according to the first principle, this quantity is pinned on its initial value, so that it is equal to U at every future time t.

It is a remarkable feat—again—that U would remain constant even though the motion is now much more complicated; imagine all the collisions that will occur when many balls are moving simultaneously on the table. But it raises a subtle hope: U has to be constant, but the individual u_n does not. All of them could vary—in fact all of them do vary. It is required only that they always add up to the same number U. In other words, they have to make up for each other: if one of them decreases, some other one has to increase. Suppose now we are not interested in all the balls that are on the table, but just in one of them, say the first one, which is black, while all the other ones are white. Would it be possible to devise a billiard table that would decrease the uncertainty on the white ball, u_1, while increasing the uncertainty on the black ones? So u_1 would decrease while $u_2, u_3, \ldots, u_N$ would increase, keeping the total uncertainty $u_1 + u_2 + \cdots + u_N$ pegged to its initial value U. We would end up knowing less about the white balls than in the beginning, but we would not care, since we are only interested in the black one, perhaps because it is the one we have to sink in a pocket.

This is a tempting way to get around the first principle: transferring information from the white balls to the black one. Unfortunately, this cannot be done. This is essentially the content of the second uncertainty principle, which was discovered by Gromov:

Second uncertainty principle in classical mechanics: Information cannot be transferred. Given the initial uncertainty regions for all N balls, there is a number r such that the uncertainty region for the black ball can never be enclosed within a circle of radius r.

Some comments are in order. First, the number r which appears in this statement depends on the initial uncertainty regions: the smaller these regions (so that the positions and velocities of every ball on the billiard are known more accurately), the smaller this number (so that better predictions can be made on the position and velocity of the black ball in the future). Gromov's principle does not tell us that there is a general limit below which our instruments will never reach. It simply tells us that, depending on the quality of our initial observations, there is a limit to the accuracy of the predictions we can make on the future behavior of the black ball. There is no way to devise a system that would indefinitely increase our knowledge of the position and velocity of that ball, while losing track of all the other ones.

It could be the case, however, that the uncertainty u_1 on the black ball decreases indefinitely, and becomes eventually smaller than any prescribed number. This does not contradict the second principle, because the number u_1, which is the area of the uncertainty region, tells nothing about its shape. We can have—in fact, we do have—regions which have very small area and which cannot be contained within a small circle. Imagine for instance a region shaped like a very thin and very long ribbon. We can make its area as small as we wish, just by making it thinner, and it can take up as much space as we want, just by stretching it. If we stretch it to length L and make it straight, for instance, then we will need a circle of radius $L/2$ to box it in. So the uncertainty region of the black ball may remain too large to be enclosed in a circle of radius r, while u_1 decreases to zero.

The first principle tells us that we cannot build a billiard table that will correct the game of a poor player. The second principle tells us that we cannot put other balls on the table so cunningly and accurately that they will correct the game of a poor player. Both principles extend to much more general situations: all systems in classical mechanics are subject to the two uncertainty principles. Both are closely related to the stationary action principle. However, the argument becomes quite technical, and we will refer the interested reader to appendix 2. Let us just conclude by pointing out how remarkable it is that the final triumph of Maupertuis' views in mechanics comes as his metaphysical views have been so utterly defeated.

(CHAPTER 6) **Pandora's Box**

THE GRANDIOSE VIEWS of Maupertuis have been laid to rest. The laws of physics, as he understood them, were striving to minimize the expenditure of a miraculous quantity, called the action. Maupertuis saw this as definite proof of intelligent design: the laws of physics were simply expressing God's purpose in his creation. Unfortunately, as we found out, the principle of least action is a misnomer; it should really be called the principle of stationary action. There go the metaphysics: there is no ready interpretation for a stationary action principle, as there was for the least action principle. Maupertuis could write eloquent volumes about this marvelous quantity, the action, which was obviously so precious that the whole order of nature was directed toward preserving it as much as possible. It is very difficult to say anything of that kind when it turns out that nature is not minimizing the quantity of action it uses, but is just trying to make it stationary. What is so important about stationary points? They are like mountain passes; they are neither high points (maxima) nor low points (minima).

The fact that light, for instance, does not take the fastest path from one point to another shows up very simply when it reflects on a mirror. Going back to figure 5, we have seen that the path *AOB* is shorter than any other path hitting the mirror, such as *AMB*, and we have concluded that the light ray going from *A* to *B* follows *AOB because* it is the shortest path. But this is clearly wrong: if the ray of light, or whoever is setting its course, were really intent on minimizing the time of travel, it would go directly from *A* to *B*, and there would be no reflection on the mirror. What if we put a screen between *A* and *B*, so that *A* cannot shine directly on *B*? If the principle of least action, as stated by Maupertuis, were true, the ray of light would take the shortest path from *A* to *B*, and that by no means reaches the mirror: it goes directly from *A* to the bottom of the screen, and up again to *B*. We know this does not happen in nature. There are two rays reaching B from A, the direct one AB, and the reflected one AOB, and that is why B sees A itself and a copy, which is its image in the mirror. If a screen is put between A and B, only the reflected ray remains, and B sees A only through the mirror. The ray AOB, which subsists in both cases, never is the shortest path.

We have seen, however, that even though the least action principle is dead, the stationary action principle is alive and well, and in the preceding chapter we have described many of its uses. So there is still a mystery to be explained: how does the light know which path to follow? How come it knows about stationary points while we don't? Do photons figure out the action along every possible path and pick the right one? In the present chapter, we shall try to explain the mystery. This will lead us to realize that the stationary action principle holds only at a certain scale. Quite different principles rule the world at the scale immediately below and immediately above.

Where does the stationary action principle come from? Can it be explained from basic physical principles, or are we to assume some mysterious sense of purpose in nature? This is the question that Clerselier asked and that Fermat evaded in 1662. Already in 1677, Huygens had an answer. Most people at the time, including such luminaries as Descartes and Newton, believed that light consists of small, hard particles traveling through empty space, and that the rays are merely the individual trajectories of these particles. Huygens, on the other hand, thought that space is not empty, but filled with an invisible medium, and that light consists of vibrations which propagate in space much as waves travel on water.

If you throw a stone into a pool, you will see circular waves emanating from the impact, propagating on the surface and being absorbed or reflected by the banks. This pattern does not arise from a global design, but from local interactions. Once the disturbance created by the stone at O has reached some point in the water, say P, it functions as a new source of disturbance, and sends waves in all directions. If we put a screen across the pond, with a hole at P, so that all the disturbances emanating from O are shut out except those which reach P, we see a new pattern of circular waves emanating from P, as if a new (smaller) stone had been thrown in at P. If the screen is removed, this pattern disappears, because it has to be superimposed upon similar patterns emanating from all the other points at which the water has started vibrating. The end result is the original pattern of circular waves emanating from the center O. It arises not because the waves sent by O travel across the water without disturbing it, but because the disturbances they create along the way cancel out except in one single direction. The waves leave a multitude of new sources behind them, but these sources interact and, at the global level, the only one we see is the initial one at O.

This is the fundamental difference between waves and particles. Particles always add up: if you put two particles into a box, there will be two more particles in that box. Waves do not always add up: if you send two waves into a box, you will have a single, more complicated, wave in that box. You may even get nothing: the two waves could just cancel out. Working out the interaction rules, Huygens found that, in the case of an initial disturbance arising at a point O, as when a stone is dropped into water, the resulting waves propagate in straight lines emanating from O. In the case of light, these would be the rays. The mathematical reason why these rays are straight lines, as it arises from computing the interactions, is that they are stationary points for the length. In many cases they are even better; they are minimum points of the length. That is, any other path with the same extremities will have greater length, although this fact is not important in itself. Only stationarity matters. We are really comparing every path to neighboring ones; if the difference in length is small enough, the path is stationary, regardless of whether that difference is positive or negative. All paths with that property qualify as rays of light. This is precisely what happens in the case of light reflecting on a mirror, where light reaches the same point by two different routes, the direct one and the reflected one.

Huygens turned out to be correct. Light does consist of vibrations, the different wavelengths corresponding to the different colors, and his explanation provides solid physical ground for the fact that rays follow stationary paths. But what about classical mechanics? Why should the stationary action principle hold for solid bodies? Surely billiard balls are not waves? The great Richard Feynman, in the mid-twentieth century, came up with a bold idea, very characteristic of his way of thinking: solid bodies pick their paths at random. This randomness is observed for very small bodies, like electrons, and the reason it is not observed for larger bodies, such as billiard balls, is due to cancellations, very similar to the cancellations which occur in wave propagation.

Think of a massive body, a small one like an electron, or a large one like a billiard ball, starting from A and ending up at B. What path did it take? The answer we get from classical physics, in the absence of any external forces, is: a straight line. Feynman's answer is: every path from A to B is possible, from the straight one to the most crooked you can imagine, but they are not equally likely. To find out how likely a given path is, one has to compute the action (yes, the classical action, as defined by Maupertuis, Euler, Lagrange, Hamilton, Jacobi, the old crowd) along that

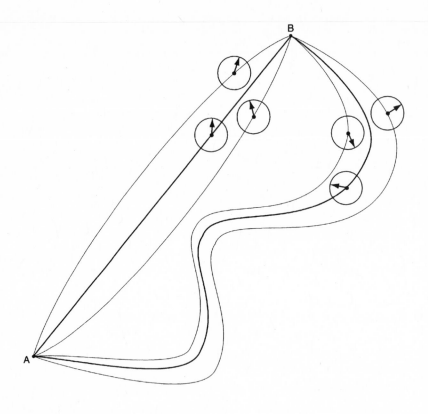

18. FEYNMAN'S PRINCIPLE According to Feynman's interpretation of the laws of subatomic physics, a particle going from *A* to *B* is not constrained to lie on the straight line *AB*, as in classical mechanics. Any curve connecting *A* and *B* is possible—but they are not all equally likely. Along each of these curves, the classical action is computed, and divided by *h*, a very small number. The result is a very large number, which is interpreted as an angle. The picture gives six of these possible paths, and the associated angles. It paths are close to the straight line *AB* (the three upper paths in the picture), the angles are almost the same. If they are not (the three lower paths), the angles are different—the difference being proportional to the mass of the particle. In Feynman's interpretation, one computes the probability of any given path by adding up the angles associated with neighboring paths: in the first case, all the contributions are in the same direction, so they add up, while in the second, they go in different directions, and so they cancel. So, for massive particles, and all macroscopic objects, the only paths which have a significant probability of occurring are the ones near the straight line. This is what happens in classical mechanics.

path. However, the probabilities do not add up, as in classical probability theory; they interfere, as in the theory of wave propagation. The most likely paths are those with the least interference from their neighbors. If one goes into the mathematics, one finds that they are precisely the ones that make the action stationary, that is, those that classical mechanics would indicate.

In Feynman's theory, all paths are possible; the classical paths are just more likely than the others. One more factor comes in: Planck's constant h, which is an extremely small number. The probability that a nonclassical path occurs, that is, that the body follows a path that does not conform to classical mechanics, behaves like h/m, where m is its mass. In other words, this probability becomes significant only when m is very small, which puts us firmly inside the subatomic range. There is no chance of ever observing anything bigger than an atom disobeying the laws of classical mechanics. On the other hand, an electron is very likely to be away from the classical paths, and there are by now numerous experiments to show it. In fact, the path of an electron cannot be predicted. The best that can be done is to compute the probabilities of the various possible paths, according to the Feynman rules. The theoretical values of these probabilities have been checked experimentally to a very high degree of accuracy.

We end up explaining one mystery by another. Clerselier's question has been answered: neither light nor stones choose the path that will make the action stationary. Massive bodies, be it atoms or stones, pick their paths at random according to certain probabilities, which can be computed before the fact. Now why it should be so is a new mystery. Why draw lots? According to Einstein's famous dictum, God does not play dice—at least, he should not be caught doing so. But what we are saying here is that the path of an electron cannot be predicted, even if one knows precisely the state of the world and has unlimited computational power; the best we can do is to compute the probabilities for the electron to go one way or the other. Beyond that we cannot go; the reason, if any, why it goes one way rather the other, is hidden from us.

This effectively kills Maupertuis' vision of the real world as the best of all possible ones. We are no longer saying that there is a quantity, called the action, which all natural motions strive to minimize, or to make stationary. We are merely saying that natural motions occur at random, according to certain probabilities. Certainly there is no idea of optimality there, no reason why the real world, that is, the set of all events that actually occur, should be better than other possible ones. When randomness rules, when events occur without definite causes, there is no meaning to be sought. If there is a God, he has left no tracks in the laws of physics; or if he has, he has covered them up very well.

Strangely enough, randomness also appears at a much higher scale, namely our own. It is a different kind of randomness, connected with chaos theory: we do not deal with events without apparent causes, as when an electron travels one path rather than another, but with events with minuscule causes, as when a rolling die falls on one side rather than the other. In this way, classical mechanics (including the stationary action principle) appear to be valid only in a thin layer of reality, trapped between the subatomic scale, which is ruled by quantum mechanics and the Feynman probabilities, and the human scale, which is ruled by thermodynamics and decaying entropy.

Indeed, at our own scale, the laws of classical mechanics are highly idealized. There is little resemblance between the idealized billiards of the preceding chapter, with the ball moving indefinitely around the table, and real billiards, were the ball cannot afford more than a few bounces before slowing down and stopping. Galileo's idealized pendulum, which swings forever, without losing energy, is a fiction. At best, under very careful and controlled conditions, one can keep the motion alive for a few hours, but eventually it is stopped by various kinds of friction on the swinging parts. In fact, it is very difficult to turn a real-life pendulum into a practical instrument for measuring time; Galileo himself did not succeed, and Huygens was the first one to build a clock on this principle.

Most differences between the laws of classical mechanics and our own experience of nature can be brought under one common heading: the irreversibility of time. Most of the events that occur around us happen in a certain order, which cannot be reversed. We grow older, and there is no way we could get younger. If we are shown two pictures of the same person, we can tell which is the most recent. Suppose we are shown two movies, showing a spoon stirring a cup of coffee. In the first one, a drop of milk which is dropped into the coffee spreads around, in the second, a brown mixture separates into a black liquid and a white drop which finally jumps out of the cup. We know very well which movie has been run backward; we also know that, in real life, once we have mixed the milk with the coffee, we can never separate them again. This irreversibility (which the late Stephen Jay Gould refers to as time's arrow) does not occur in classical mechanics. Nothing is easier than to reverse the motion of a billiard ball on the idealized tables we have been playing on: just send the ball back along the direction it came from, and it will nicely retrace its whole trajectory.

Here we have a new mystery: how can we reconcile our basic experience of life, which is that clocks do not turn back, with the laws of classical mechanics, including the stationary action principle, which make no such distinction?

To make the contradiction more apparent, let us perform a thought experiment; call it Pandora's box. Pandora has been given a hermetically sealed box, under strict orders not to open it. She does not know it, but the box contains a certain very rare gas, called imaginum. Imaginum, like all other gases, consists of a huge number of molecules with different speeds, moving about, colliding with themselves and bouncing off the walls. Pandora, curious to know what is in the box, opens it; to her dismay, the imaginum, which is a nice blue color, immediately flows out. She jumps to the door and closes it, so that now the room is sealed, and she starts wondering how to get the imaginum back into the box.

Just to make things simpler, suppose there is no air in the room, so that the imaginum just flows out into empty space. Then there will actually be some imaginum left in Pandora's box: the flow stops when the pressure in the box and in the room are equalized. If we look at what is happening to the molecules, those which are near the top when Pandora opens the box have a good chance to be pushed away by the others, and they will run into the room. Once they are there, it is not likely that they will go back into the box, since there are no molecules in the room to push them back. More molecules will follow them, and the whole process will go on until the pressures in the box and in the room are equal, that is, until the molecules on top of the box are equally likely to be pushed up into the room or down into the box.

From then on, there will be equilibrium; that is, there will be no further changes. Molecules will still move around, sometimes with great speed, and they will find their way from the box to the room and vice versa, but the pressure of the imaginum will remain constant, because it is a statistical average.

At least, that is what common sense tells us. There is no way Pandora can get the imaginum back into the box. What she has done cannot be undone, and she has to prepare herself for the consequences; this is exactly what time's arrow is all about. However, mathematics tells us a different story. According to a famous theorem by Poincaré, the imaginum will eventually go back into the box by itself. All that Pandora has to do is to be patient enough, wait until all of the blue gas is back into the box, and then snap the lid shut. There could be no clearer contradiction. On the one side, irreversible time: the past is lost forever. On the other, cyclical time: eventually, things revert to what they were. What is the basis for such a claim?

Although Poincaré's theorem runs counter to our intuition, the reason why it is true is easy to understand. Imagine that, to start with, the pressure in the box is really very low, so low in fact that there are just a few molecules knocking around. Suppose first there is only one, and the room is quite small; after leaving the box, the molecule will wander aimlessly, exploring every corner in the room, so aimlessly in fact that eventually it will find its way back into the box. This is a bit like a drunkard walking at random and knocking at every door he goes by: he will eventually find his way back home. Now increase the number of molecules. Suppose there are ten knocking about in the box when Pandora opens it. They all start going back and forth, like the first one did; they all spend some time back in the box; and it is to be expected that at some future moment they will all be together again. It is also to be expected that, at some other moment, there will be just one of them in the room while nine are back in the box. In fact, anything can happen, and will happen if one just waits long enough. Now, instead of ten molecules, let us put one hundred, then one thousand, increasing until we reach 10^{23}, one followed by twenty-three zeroes. This is about how many molecules a half liter of air would contain in normal conditions. The same arguments hold: it is to be expected that, eventually, all the molecules will be back in the box. So Pandora just has to be patient.

The key to the paradox, of course, is the time involved. To see the imaginum flow back, Pandora must be prepared to wait much, much longer than the expected duration of the universe. In any shorter time, in anything that can be counted in mere billions of years, nothing like this is remotely likely to happen. For mathematicians, of course, this makes no difference, but for human beings, especially for Pandora, who will have some explaining to do rather soon, it does. Poincaré's theorem is true, but it does not help us. Time's arrow appears at our scale because the objects we deal with are large aggregates, and they will disappear long before they can exhibit any tendency to retrace their past history.

So this is the first source of irreversibility. There is one more. Let us play billiards once more, this time with three balls. As we pointed out earlier, this system apparently is insensitive to time's arrow. If we want the three balls to retrace their paths, all one has to do is to reverse their velocities and send them back along the directions they came from. They should then retrace their steps, that is, travel backward in time: if nothing stops them, one minute from now they will be exactly where they were one minute ago.

This is not the case: there is no way one can retrace the paths the balls followed, or find out where they were one minute ago. The key to the mystery lies in the word "exactly": it belongs to mathematics, not to physics. No physicist can guarantee that two quantities are "exactly" equal. All the values he gives are the results of measurements, and they all come with a margin for error, which depends on the instruments he uses. It is not feasible to send back the billiard balls exactly in the directions they came from, with exactly the same speeds. The best that can be done is to send them back almost in the directions they came from, with almost the same speeds. Is that good enough?

The answer is no, because of a special feature of billiards, which it shares with many other systems, and which is at the core of chaos theory. Small initial errors get amplified very quickly with time, mainly because of the collisions between balls. As a result, any maladjustment we made in sending the balls back (there is bound to be one, and even if by extreme luck there were none, how would we know?) will veer it off course, so that the trajectories we observe after a few collisions will have nothing to do with the original ones. In fact, the system is so sensitive that even Newtonian attraction will influence it. Surprising as it may be, already after nine collisions, the presence of people around the table will modify the trajectories. If we wanted to go beyond that, and to follow trajectories for one full minute, we should have to take into account people walking in the streets or airplanes flying in the air. This means that there is no hope of ever reconstructing past trajectories: for all practical purposes, the system is as random as a game of dice.

In both examples, Pandora's box or the billiards, it is still true that the present state of the system contains all the information necessary to reconstruct its past history and predict its future trajectory. However, this information is irretrievable, and this is what creates time's arrow.

As another example, consider the so-called baker's transformation. Start with a sheet of dough, knead it down with a rolling pin to about half its height, and fold it back over itself. We now get a new sheet, with the same height as the first one, but consisting of two layers, the right half above the left one. We now incorporate dark chocolate into the upper layer, so that it is now black while the lower one remains white. Knead it down again, cut in half, and stack one half on top of the other. We get a four-layered sheet, with black and white alternating. Do it again, and again, and again. After n steps, we get a sheet of the same height as the original one, with 2^n layers of alternating colors.

Note that this transformation is reversible. One can peel off the sheet and restore the original height by putting the top layer beside the bottom one.; repeating the procedure enough times, 2^n to be exact, will restore the sheet to its original state (with its right half black and the left one

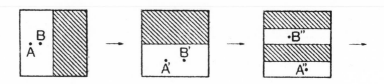

19. THE BAKER'S TRANSFORMATION These are the three initial stages. The square on the left is flattened (its height is divided by two while its width is multiplied by two), yielding a rectangle. The rectangle is cut in two and its right side is put on top of its left side. This yields the square in the middle. The same procedure is repeated, flattening, cutting and putting on top, and this yields the third square. The points A and B are brought to A' and B' by the first step, and to A″ and B″ by the second. As the transformation proceeds, the white and black stripes become thinner.

white). Better still, one can compute the original position of any point in the sheet from its position at the nth step. For instance, if a point lies in the lowest sheet (white), at a distance d from the left side, its original position was at a distance $d/2^n$ from the left side. If it lies in the uppermost sheet (black), at a distance d from the right side, its original position was at a distance $d/2^n$ from the left side.

If a point does not lie on the upper or lower side of the sheet, but inside the dough, the question becomes more difficult to answer. We first have to find out whether it lies in a black layer or in a white one; in the first case, its original position must have been in the right side of the original sheet, in the second it must have been in the left side. But by now, after n steps, the layers are 2^n times thinner than the original sheet; if $n = 10$ the height has been divided by one thousand, and if $n = 20$ by one million. Certainly, by $n = 10$ we will not be able to distinguish the layers with the naked eye, all we will see is a uniform grey. By $n = 20$, we will be worrying about the limits of the instruments we are using to measure the positions, and by $n = 30$ we will be way beyond them. By this time, even though the information is there, we will be unable to retrieve it.

In theory, no information has been lost along the way, since the system is reversible. In practice, the information is too fragmented to be recovered. The system dilutes it progressively until if falls below the threshold of observation. Once this has happened, the best we can do is to say that any point in the dough has a 50 percent chance of being black, and a 50 percent chance of being white. Randomness raises its head again, but it is of a different kind than the one we found at the subatomic level. This time, it arises not because there is a source of randomness somewhere in the system, but because the power of our instruments is limited. To distinguish it from the first, we will call it *chaos*.

Chaos cuts with two edges. We have seen how it is impossible to retrieve past history from current observations. We will now show that is impossible to predict future states from current observations. Indeed, suppose we apply the baker's transformation a large number of times, say n, and we want to know whether a point which started off somewhere inside the sheet will end up in the lower or upper half. Denote by M_1 the initial position, and by M_n the position after n steps. Remember that, at each step, the dough is stretched to twice its length and cut and stacked, the right half above the left one. If one works through the mathematics, one finds that, to answer the question, one has to know M_1 with an accuracy of about $1/2^n$; if one makes an error of that size in locating M_1 horizontally (its height does not matter), one will end up on the wrong side. Again, the precision of our instruments is limited, so that for large n it is impossible to know the initial position with the required accuracy. The best we can do is to say that there is a 50 percent chance of M_n ending up on the top, and a 50 percent chance of its ending up in the bottom.

We have lost a great deal, but we are not entirely empty-handed, for this kind of probabilistic prediction can be greatly generalized. Suppose we are interested not in the top and bottom half, but in the left and right half of the sheet. The chances of ending up in either, after a large number of steps, are also 50 percent: they do not depend on the shape of the halves, but only on the fact that they are halves; that is, they occupy 50 percent of the total volume. Suppose now we divide the sheet into two parts, a large one, A, occupying 99 percent of the volume, and a small one, B; then the chance of ending up in A is 99 percent. At that stage, we can predict, with some amount of confidence, that the final state M_n will lie in A. This is by no means certain, for M_n may well end up in B. But the probability of its doing so is less that 1 percent, so it is not very likely. Predicting that the final state will lie in A is a safe bet, and it becomes even safer if the size of A increases. If for instance A occupies 99.99 percent of the total volume, then there is only one chance in ten thousand of our prediction failing, and it becomes an almost sure bet.

There are situations where the probability of our bet's failing is zero, or so low as to be zero for all practical purposes. This is what happens in the case of Pandora's box. Once pressure between the box and the room has equalized, will the imaginum flow back from the room into the box? In other words, can the gas spontaneously go from a situation where it is evenly spread around to a situation where it is concentrated in a small region of available space?

Call A the first situation. From the point of view of the molecules, A covers a lot of different cases. Saying that pressure is equal in the box and in the room tells you that there are about the same number of molecules in each, and that the distribution of velocities is about the same, but it does not tell you which molecule is where and what is its velocity. There are many ways to achieve this global situation; for instance, it makes no difference to the observer that a certain molecule of imaginum, rather than another one, is in the box. One can look at all the possibilities and derive a probability for A.

It turns out that this probability is so close to 1 that we can safely bet that situation A will prevail forever. If we did so, and someone bet against us, he would have to wait billions of billions of years before observing anything that would give him even some hope of winning. We should not bet with God, for he is patient, and can afford to continue waiting after this universe ends, but with mere humans we are perfectly safe. It is a practical certainty that imaginum will never flow back into Pandora's box. Time is irreversible at our scale.

We are more or less at the end of the road. We have found randomness at the subatomic scale, and chaos at our own scale, with the stationary action principle caught somewhere in the middle. Nothing much is left of Maupertuis' dream. If not in physics, where are we to find the best of all possible worlds? Perhaps in biology? Let's give it a try.

(CHAPTER 7) May the Best One Win

STATEMENTS LIKE "This world is the best of all possible ones" or "This student is the best in his class" have the same logical structure: they are comparisons between a given object or person and others. For such sentences to be meaningful, one must define the class of objects or persons with which the given one is being compared, and the criterion of "goodness" which is being used to rank them. For instance, there are many ways to rank students. Some may be better in music or science, others in English or athletics; John may have been the best one last year, but may now be falling behind. Different criteria give different rankings, and unless Jack consistently outshines Jill in every field of human endeavor, it is always possible to find a criterion which will put Jill ahead of Jack, and still another one which will put Jack ahead of Jill. Stating that "Jill is the best of her class" means that a certain weighted average of all grades has been defined, and that, according to that criterion, Jill comes out on top.

These statements translate mathematically into an *optimization problem*: a certain class (of objects, persons, situations, and so forth) has been delineated, and a numerical criterion (for instance a grade) has been defined for ranking its members. Solving the problem means finding the top-ranked element (in mathematical terms, *maximizing* the criterion), which is then referred to as the *optimal* one, or simply the *optimum*. In some cases, one would rather minimize the criterion instead of maximizing it, that is, find the least-ranked element instead of the top-ranked one, but this is of no consequence, and one will still call the result optimal.

Maupertuis' fundamental intuition is that our world is optimal: it is the one which uses up the least quantity of action. The class of comparison consists of all "possible" worlds, and the criterion is the quantity of action, which turns out to be minimized instead of maximized. In the preceding chapters, we showed that Maupertuis was mistaken. The quantity of action is not minimized, nor is it maximized. In addition, the laws of nature are different at different scales, and if there is a unifying principle to be found, which I personally find very unlikely, we have no idea what it will be, and there is no reason why it should turn out as an optimization problem.

On the other hand, there is a definite sense of progress in the way we think about the world, if not in the physical sphere, at least in the biological and historical spheres. We often think of humanity as rising from the dark ages, when tribes of hunter-gatherers were eking out a miserable subsistence from a hostile environment, to the affluent societies which arose from the industrial revolution. Moving farther on in time, it is hard not to qualify as progress the evolution from the first anthropoids to modern man, or from unicellular forms of life to multicellular ones. In fact, the history of life on Earth has often been seen as directed toward bearing forth its most perfect form, namely ourselves. The tree of evolution is then pictured as a ladder, rising from inert matter to more and more complex forms of life, from bacteria to humans, standing on the last step. It is also quite common to think of evolution as continuing beyond humans toward higher and better forms of life, perhaps even achieving some kind of divinity; although the latter idea has become unfashionable nowadays, it is quite a natural way to reconcile the basic requirement of Christian theology (our salvation) with the basic fact of biology (evolution).

The vision of evolution driving progress is a dynamic version of Maupertuis' vision of God creating a perfect world once and for all. It is no longer claimed that the world is the best of all possible ones; rather, it is getting better every day: it is not optimal, but it is continuously improving. The process at work is no longer the benevolence of a rational Creator, but the blind forces of evolution. All living creatures are permanently engaged in a struggle for scarce resources, such as space, sunlight, food, and sexual partners while fighting off enemies such as parasites and predators, or in the case of humans, other members of their own species. This "struggle for life" results in the "survival of the fittest," while the losers are to be sought in the fossil, prehistoric, or archeological records.

One way to think about this situation is to see it as an optimization problem, where the criterion is "fitness." By definition, fitter is better, and nature, by letting only the fitter ones survive, solves an optimization problem. As a logical consequence, whatever survives has shown itself, by the mere fact of still being around, to be better than what it has replaced. Transposing this idea into history, one gets arguments to prove that Neanderthal man was an inferior species, that the European settlers were better than the various American, African, or Asian people they exterminated, displaced, or enslaved, and more generally that might is right. This is the basic claim to superiority of Western civilization: we have an overwhelming military advantage, which enables us to take land and resources away from others and put it to our own use, so our way of life must be the best one. The white man's burden is to bring Western civilization to all peoples, not to the extent, however, of letting them have the same weapons that we do, for that would be surrendering our moral superiority.

The optimization approach is wrong. On the one hand, "fitter" means more adept at survival, but not "better" in any reasonable sense: more complex, more intelligent, or more moralt. The great success story of evolution are the bacteria, which have shown their ability to survive over 3.5 billion years, with very little change in the basic blueprint, and even today are populating the most extreme environments on Earth, such as the thermal vents on the ocean bed, where some species prosper in temperatures as high as 400°C and pressures of up to 300 atmospheres. On the other hand, fitness is not a criterion which applies to the world as a whole, and which would allow us, for instance, to compare the Jurassic world, where dinosaurs roamed the primitive continents, with the present one, which is almost entirely taken over by humanst. Fitness is a relative criterion, relating a species, an individual, or a gene to its environment, that is, to the world it lives in. If the environment changes, the fitness criterion changes; polar bears are certainly fitter than rattlesnakes for life in the Arctic, while the opposite is true in the Mojave Desert. Dinosaurs and people cannot be compared; one cannot be declared fitter than the other, because they never shared the same environment.

In Darwin's own words, the way the struggle for life operates is "descent with modification"; the word "evolution" is not his, but Spencer's, and Darwin never liked it. Each generation of individuals produces another, much more numerous in general than would be required for preserving the species if the survival of each offspring was guaranteed. All of the new individuals are born different, from each other and from their parents, and will transmit the modifications to their own offspring.[1] They immediately fall into a hostile environment, and those who have benefited from favorable modifications have an edge in the struggle for life. It is a probabilistic effect: the advantage may be slight, and make little difference for any given individual, but accumulated over many

---

1. Nowadays, we have material support for this assertion that Darwin did not have, namely random mutation in the genetic material.

individuals and many generations, it is enough to shift the whole species in one direction, or even result in a new species. This does not mean that the new species will be better than the old one in any absolute sense; it is just better adapted to its environment. Note that it may be a new environment, different from the one the original species lived in; the climate may have changed, or the mutation may have occurred in individuals who had migrated.

The case of the Galapagos finches is famous: Darwin noticed that there were many different species of these birds, all with different beaks—short and sturdy ones to crack nuts; thin, needle-like ones to suck blood from larger animals—and he concluded that they all descended from a single species of mainland finches, which had adapted to the different sources of food available on the islands. Parasites provide another instance of extremely specialized adaptation. The tapeworm, hooked into the intestines, deprived of any organ for locomotion or perception, is well fitted to its environment, but it would not survive under any other conditions. In fact, parasites are so helpless outside their host that one can hardly see them as winners in the struggle for life, unless one understands that the world the parasite evolves in is its host, and that it cares as little about other possible worlds as we do.

These are simple situations: finches competing for stable sources of food, or a parasite adapting to its host. Most species occupy ecological niches in a food web, which is a complex system of predator-prey relationships. We now enter a very complex situation, for the environment of each species is constituted by all the others. In other words, the world against which the fitness of species A has to be measured is nothing but the set of all species alive at that time, including A itself. The process of descent with modification is supposed to change species A so as to improve its fitness against that particular world, that is, against species B, C, and all the others. But the same process is simultaneously at work in species B, and changing it as well, not to mention species C, D, and all the others. As a result, all the species are evolving together, so that every species finds that its environment is changing, and has to adapt itself to the new conditions. This is a much more complicated process than straightforward optimization, and one may well ask what the overall effect might be.

Off the West Coast of South Africa, the waters of Malgas Island are dominated by seaweeds and rock lobsters that prey on mussels and whelks. Nearby Marcus Island is similar in every respect, but its waters have extensive mussel beds and whelks at high density, lobsters and seaweed being notably absent. Local fishermen relate that, around 1965, there were lobsters on both islands; so, in 1988, they tried to reintroduce lobsters on Marcus Island. In a famous experiment, one thousand lobsters were transferred from Malgas to Marcus.[2] To the experimenters' amazement, the whelks overwhelmed the much larger creatures, crowding on them and eating them up, so that within a week there was not a single lobster left. The lesson to be learned from this example is that there is a feedback loop between the individual species and the global environment: on the one hand, the environment determines the behavior and evolution of every species; on the other, the environment is nothing but the set of all species living together in that ecological system. Even such a basic relationship as who preys on whom depends on the environment: on Malgas, lobsters prey on whelks; on Marcus, whelks prey on lobsters.

This drives home our basic point: fitness is a relative thing, and survival of the fittest does not lead to any kind of overall optimum. Lobsters are on top of the food chain on one island, and on the other they are unfit for survival. Malgas and Marcus offer two different biological answers to the same set of geological and geographical conditions. It is hard to see why one should be called better than the other; certainly, fitness to the environment does not provide such a criterion. Darwin himself was well aware of this, and points it out with his usual care in the *Origin of Species:* "Natural selection tends only to make each organic being as perfect as, or slightly more perfect than, the other inhabitants of the same country with which it has to struggle for existence. And we see that this is the degree of perfection that is attained under nature. The endemic productions of New Zealand, for instance, are perfect compared to one another; but they are now rapidly yielding before the advancing legions of plants and animals imported from Europe."[3]

So Maupertuis' great vision finds no more support in biology than in physics. It is not the case that evolution drives the world toward an optimum. The best the Darwinian process of descent with modification can do is to lead an ecological system to some kind of equilibrium, where every species is adapted to all the others it lives with. Such equilibriums are complex, and depend on all the species that interact: once the lobsters have disappeared from one island, as was the case on Marcus, there is no reintroducing them. The old equilibrium has been destroyed, and the ecological system has found a new one, where there is no room for lobsters.

2. A. Barkai and C. McQuaid, "Predator-prey role reversal," *Science* 242 (4875) (October 1988): 62–64.
3. *The Origin of Species* (London: John Murray, 1859), chap. 6.

It would also be wrong to imagine that the struggle for life will ever end in some kind of general equilibrium. The past history of life on Earth seems to have been driven by random events, which jolt the environment out of the state it had settled in and start afresh the whole process of descent with modification. In the case of Marcus, the original reason why the lobsters disappeared, be it overfishing, sickness, or some other natural disaster, is not known, but it certainly switched its biological future onto another track, while Malgas stayed on the same path. At a much larger scale, 65 million years ago, the dinosaurs disappeared from the Earth and left it free for mammals to occupy. Another mass extinction, 225 million years ago, wiped out 96 percent of all species living in the oceans at that time. It is thought nowadays that such mass extinctions were triggered by catastrophic changes in the environment, due to a large meteor hitting the Earth, gigantic volcanic eruptions, global warming or cooling.

This puts randomness squarely at the center of the picture. On the one hand, our future seems to depend on a mass of objects orbiting the Sun, some of which may end up falling down on us, or on largely unknown processes unwinding miles below our feet, and which may result in huge amounts of magma seeping out of faults in the Earth's crust. On the other hand, if such an event happens (or rather when, for it is bound to happen sooner or later), the final outcome will be no less random than the process itself, for the catastrophe will quite likely burst out upon a totally unprepared world. Millions of years of descent with modification will have resulted in an equilibrium where species are well adapted to conditions which have very little to do with the ones that suddenly prevail. Suppose, for instance, that a gigantic dust cloud, blown up by a meteoric impact or a volcanic eruption, shuts out sunlight for several years; most existing species will not be prepared for such an ordeal. It would be as if two football teams were suddenly called upon to play water polo: it would be very hard to predict the outcome of such a contest, based on the players' previous performances. It would be decided by the ability to swim, which is so irrelevant to the basic training of professional footballers as not even to be recorded, but which would suddenly become much more important than the ability to run. Likewise, the fact that diatomeae, for instance, have survived the mass extinction of 65 million years ago, while other algae species did not, is attributed to their ability to change into spores, a quiescent and hardy form of life which may have been developed to resist seasonal fluctuations in their food supply, and which may have enabled them to survive the great night that extended over Earth.

The major events in the history of life on Earth have not all been mass extinctions. As Stephen J. Gould was fond of recalling, there was also a mass explosion, 570 million years ago, which gave birth to the first multicellular animals with hard body parts. The Burgess shale remains as a witness to this remarkable period, and has preserved a few of the species that appeared at that time.[4] Eight of them do not belong to any of the animal phyla alive today—sponges, corals, annelids, arthropods, mollusks, echinoderms, and chordates, the latter including vertebrates. There are arthropods as well in the Burgess shale, but most of them cannot be classified in the four great groups we know today, three still living and one, the trilobites, exclusively fossil. There is an astonishing creativity in the Burgess fauna, as if life had been trying out the widest possible range of forms, and left it to the evolutionary process to sort out. There were no more similar episodes; the Cambrian explosion was the first and last of its kind. After each mass extinction, the ecological niches which were freed were occupied by species descending from the survivors. This means that the old blueprints were used: there was no more room for experimenting with new ones.

As Gould points out in the epilogue of his book *Wonderful Life*, there is a single chordate in the Burgess fauna: it is *Pikaia*, "a laterally compressed ribbon-shaped creature two inches in length."[5] The Burgess fauna is teeming with life, and there is nothing in this single modest individual to set it apart from the others. And yet, this is the blueprint for most animals we can name, fishes, birds, and mammals, not to mention ourselves. Why did this particular blueprint survive until now, while most of the other species and phyla in the Burgess fauna disappeared long ago? Most probably, there is no single compelling reason, and the answer combines several degrees of randomness, from the impact of large-scale events such as mass extinctions or continental drift, to genetic mutations giving certain species at certain times an overwhelming advantage in the struggle for life. As Gould puts it, "The survival of *Pikaia* was a contingency of 'just history.' I do not think any 'higher' answer can be given, and I cannot imagine that any resolution could be more fascinating. We are the offspring of history, and must establish our own paths in this most diverse and interesting of conceivable universes—one indifferent to our suffering, and therefore offering us maximal freedom to thrive, or to fail, in our own chosen way."[6]

4. *A Wonderful Life* (New York: W. W. Norton, 1989), chap. 3.
5. Ibid., epilogue.
6. Ibid.

All this holds true in human history as well. Of course, the time span is much shorter. Recorded human history does not stretch farther than 4000 BC, when writing was discovered in Mesopotamia. Between the early Sumerians and the present time, when the last remnants of their activity are being destroyed by looters in the wake of the 2003 invasion, there are no more than six millennia, a bare flicker in the 4.5 billion years of geological time. But at this much smaller scale, we find that human history mirrors the history of life. Human societies are engaged in a perpetual struggle for space and resources, as are species. The agent of change is no longer descent with modification, but the cunning of human beings, though the playing field is largely shaped by events outside their control. Volcanic eruptions, floods and droughts, parasites and plagues, have all taken their toll.

There have been many attempts toward a scientific theory of history, comparable to the theory of evolution. In my opinion, the ones who have come closest to such a theory are Thucydides (460 BC–395 BC) and Francesco Guicciardini (1483–1540). The first wrote the history of the Peloponnesian War, which pitted the sea empire of Athens against cities of mainland Greece, led by Sparta. The war began in 431 BC and ended twenty-seven years later with the defeat of Athens, but Thucydides' account covers only the first twenty-one years. The second wrote the history of the wars which ravaged Italy between 1492 and 1534, and which ended with the Hapsburg emperor Charles V beating the French for supremacy over the peninsula. Thucydides was an Athenian, Guicciardini from Florence, both of them held high positions during the wars, and they both saw the war end in disaster for their own side. Their beloved cities, Athens and Florence, had to submit to a foreign power, which imposed on it a new constitution; along with their independence they lost their traditional freedoms.

Both their accounts are structured in the same way. At certain times, a crossroads is reached in the course of events. An important decision is to be taken: should Sparta declare war on Athens? Should Venice grant the emperor and his army free passage from Austria into Italy, through its territory? The decision-making body assembles, one person rises and advocates one course of action, with convincing arguments, then another person rises and advocates the opposite, with equally compelling arguments. The assembly then chooses which advice to follow, and the action runs its eventful course again, on a road which usually turns out to be full of accidents and surprises, until a new crossroads is reached. The consequences of the decision can be very far from what its advocates forecast, either because the original advice was bad, or because unexpected events threw the action off course.

As a typical instance, consider the beginning of the Peloponnesian War, as Thucydides describes it.[7] The Corinthians came to Sparta to complain of the many encroachments by the Athenians, and urged the city to declare war. Their basic argument was that it was now or never: every day the Athenians were chipping away at Sparta's allies, and if they waited any longer, they would find themselves isolated and facing a much stronger enemy. Then Archidamus, king of Sparta, spoke against declaring war. He argued that this was a war that could not be won. Athens drew its power and prosperity from the sea, and most of its allies lay on islands or in Asia Minor, on the other side of the Aegean. Spartans had no fleet. There was no hope of taking Athens by force, since the city was fortified, nor by starvation, since everything it needed came through the harbor. The best they could do was to plunder the neighboring territory, which might inconvenience some landowners but would not bring the city to its knees. It is much better, said Archidamus, to keep our options open, and try to contain the Athenians by diplomacy, while building up our strength.

In fact, his advice was not followed, and the Spartans declared war and invaded Athenian territory. Everything then followed as he had predicted. The Athenians did not fight the invasion, but retired behind their walls, from which they watched the Spartan forces plundering the fields and burning the houses.

Meanwhile it was business as usual: ships went in and out of the harbor, bringing into the city food and silver, and carrying troops on military expeditions. Almost every year of this long war saw the return of the Spartan forces, destroying the crops around Athens, until the countryside finally lay deserted. Meanwhile the real war was being waged elsewhere, Athens was assembling more power and wealth by subduing more islands, until a wholly unexpected event took place: the great plague of Athens.

Athens was a very large city for its time, and it was certainly a health hazard to have so many people, citizens and refugees alike, crowded within the walls, but an epidemic of this virulence and magnitude was unheard of at the time. It is estimated that one-third of the population died of the plague, and even today, it is not clear precisely what disease it was. The plague dealt a terrible blow to Athens and hit much harder than anything the Spartans and their allies could have done. It is certainly the main reason why the ultimate outcome of the war was not

7. Thucydides, *History of the Peloponnesian War*, trans. Charles Forster Smith, Loeb Classical Library (Cambridge: Harvard University Press, 1956–1959), 1.67–87.

what Archidamus, sound as his advice was, had predicted. Another reason was the ill-fated expedition which Athens sent to Sicily and which was utterly destroyed. Here again, an unforeseen event played a crucial role. After suffering several defeats around Syracuse, the remaining Athenian forces planned to sail home, but on the night scheduled for departure there was an eclipse of the Moon, so that the priests ordered the troops to stay for twenty-seven more days in order to placate the gods. This gave the enemy plenty of time to get reinforcements. The Athenian fleet was destroyed, the troops captured and left to rot in the quarries.

Many great failures and many great successes are due to chance and not to human folly or ingenuity. In his *Ricordi*, the notes he kept throughout his life, Guicciardini writes, "Pray God always to be found near to victory, for you will be given credit even if you have had no part in it; whereas whoever is found near to defeat is accused of infinitely many things of which he is totally innocent."[8] He also stresses the importance of another kind of chance, due not to events which lie outside human control, but to the unexpected consequences of actions which are below the threshold of attention: "Small events that would hardly be noticed are often responsible for great ruins or successes: and this is why it is well advised to consider and weigh every circumstance, no matter how small."[9] Do we not hear an echo of chaos theory? How would the 2000 presidential election in the United States have turned out if there had not been butterfly ballots in Florida, or if the voting machines had functioned properly?

Chance, of course, is not everything: there is also human decision-making. As an example, let us consider the situation the Venetians found themselves in when, in 1507, the Hapsburg emperor Maximilian asked to be allowed free passage with his troops through their territory. The Venetians were at that time allied with the king of France, and it was clearly the emperor's objective to attack the king in northern Italy, once his army had safely crossed the Alps. When all was said and done, there were two courses of action open for the Venetians: to refuse the request, and run the risk of the emperor seeking an alliance with the king against them, or to reverse alliances and join the emperor in fighting the king.

In the first of the two speeches that are delivered in the Venetian Senate on the occasion, an important point was made: what they actually wanted to do was not as important as what the others believed they

8. *Ricordi* (1512–1530), ed. G. Masi (Milan: Mursia, 1994), 176.
9. Ibid., 82.

wanted to do. Even though they wanted to be faithful to the alliance with France, the king might believe that they wanted to switch, either because he himself was treacherous and judged others to be like him, or because he suspected that the emperor was making the Venetians a more advantageous offer than he actually was. If that was his belief, he would find himself better off seeking an alliance with the emperor, and in sharing with him the spoils of the likely defeat of the Venetians. Worse still, even if the king believed that they wanted to remain faithful, as indeed they did, he might think that they suspected him of being in doubt and of preparing this kind of preemptive strike, so that they might throw themselves into the arms of the emperor out of suspicion and not of greed. In the end, the only safe way to proceed was to accede to the emperor's demand, for if they did not, the king would proceed as if they had, and they might as well reap the benefits for their own safety.

This was really a very modern analysis. It shows what an important role beliefs play in conflict situations. Of particular importance are the beliefs about other people's beliefs, and since these can never be fully ascertained, the whole situation is driven by mutual suspicion. In certain circumstances, it may even have a stabilizing effect. This was the case in Italy until the death of Lorenzo the Magnificent in 1492; the peninsula was until then divided into five major states, of about equal importance. None was powerful enough to prevail against the others, and all understood that ganging up against one of them would create a dangerous precedent which would imperil their own existence. In Guicciardini's words, "Everyone was watching the actions of the others carefully, checking any move that would have enabled one of them to increase his power or reputation; this did not make the peace less assured, but rather made all of them eager to put out immediately any small spark which could have started a new fire." This carefully crafted equilibrium was finally destroyed when the Duke of Milan called the French into Italy, the first of the many incursions which would ruin the country over the following forty years.

We now have a formal model for these situations. It is called *game theory*. The mathematical foundations were laid by John von Neumann and John Nash around 1950, and in the remainder of the century it has proved itself to be a versatile tool for analyzing economic and social situations. The model consists of individuals or groups, called agents, or players, each of whom has to decide on one course of action. Once all decisions have been made, a global situation results, which affects every agent in a different way. The problem each player faces is that he wants to get the best possible situation for himself, while knowing full well that the final outcome will depend not only on what he does, but also on what the others do. This makes it very different from simple optimization, where the outcome would depend on the agent's own actions only. A situation which is optimal from Jack's point of view may be very bad or very good from Jill's; in the first case, Jack and Jill will try to outguess each other, in the second they will try to coordinate their actions. This is strategic behavior, and we need a new concept to account for it.

An *equilibrium* is a situation where each agent's actions turn out to be the best reply to everyone else's. It is a situation of stable mutual adjustment: everyone anticipates everyone else's behavior, and all these anticipations turn out to be correct. In other words, it is a set of self-fulfilling prophecies that players formulate about each other's actions. Such situations are central to social life, because they are the only stable ones. Not to be in equilibrium means that some anticipations turn out to be wrong, so that some actions turn out to be inappropriate to the actual situation. This will lead the concerned individuals or groups to revise their anticipations and adjust their actions, thereby creating new discrepancies to be corrected at the next stage, so that the whole situation is destabilized, and the system starts oscillating wildly. In equilibrium, on the other hand, all anticipations are confirmed by experience, and every acquired behavior turns out to be appropriate in every situation, so that they become more ingrained as time goes by, and eventually solidify into social norms.

Basic features of social organizations, such as trust, or power, simply express some underlying equilibrium. Power is nothing but the illusion of power, the universally held belief that a certain person will be obeyed, that certain orders will be followed. It is self-fulfilling, for if I am given an order by such a person, I will follow it for the simple reason that if I don't, someone else will, and it will probably be worse for me. Trust is the belief that others will comply to certain rules, and every time I myself comply to those rules I strengthen the general feeling of trust. Note that distrust is self-supporting as well. If I distrust you, and you distrust me, I will take every due precaution to protect myself against your anticipated behavior, giving you good reason to distrust me a little bit more. Trust is an equilibrium, distrust is another. In the first one, everyone trusts everyone else, and is right to do so; in the second one everyone distrusts everyone else, and is right to do so. A situation where some are trustful and the others take advantage of them would not be stable, for one kind would learn from the other. Either the crooks will mend their ways and enjoy the many advantages of a society where everyone is trustworthy, or the others will be taught the hard way not to be gullible, and trust will disappear as a facilitator of social interchange.

Many rules which we assume to be universal are in fact relative to some equilibrium. If the lobsters on Malgas Island were able to think, they would believe it a basic rule of nature that lobsters prey on whelks, whereas just the opposite is the case one island away. This illusion preys on humans as well: we are born and bred into an equilibrium, the real extent of which we do not know, but which we tend to think of as the only natural or reasonable one. Think for example of the emancipation of women. In a society where women are confined to a domestic role, while public life is taken over by men, it is easy to think of this separation as being due to some inherent differences between men and women, rather than to some temporary organization of society. Acting upon this belief, one will educate girls differently from boys, so as to prepare the former for domestic chores and the latter for public roles, so that indeed they become different as adults, all satisfied with the characters they have been trained to play. It is an equilibrium, and it is very hard to break out of it; in fact, the emancipation of women in our society has been a slow process, still going on, which requires not only creating opportunities for women, but also changing minds through education.

An equilibrium is not always an optimum; it might not even be good. This may be the most important discovery of game theory.

Imagine, for instance, that there is a common task to be accomplished, and that every member of the group may either cooperate in the effort or shirk. This is the choice we face when we develop an interest in social or political issues: we can either do the hard work of lobbying for them, by showing up at meetings and taking part in the chores of organization, or we can simply skip it and wait for others to do the work for us. Let us put in some cash values. Say the group is lobbying for a tax rebate. Participating in its lobbying effort entails a personal cost of $11. If n members of the group participate, the tax rebate will be $$n$ for everyone in the group, whether they have participated or shirked. Each time a new person participates, everyone receives $1 more. This means that participating entails a personal net cost of $10, while shirking gives a free ride on future benefits. The tension is between a small benefit which accrues to everyone against a large cost to the individual who chooses to participate.

Say there are one hundred members in the group. If everyone participates, each member of the group will pay $11 herself and receive a $100 tax rebate. This looks like an excellent opportunity of earning $89, and the total earnings of the group are then $8,900: let's just go and do it! The problem is that this requires everyone to cooperate, whereas there are additional rewards for shirking. If I decide that I had rather keep my $11, and I am the only one to do so, my earnings jump to $99, while the earnings of the others drop to $88 each. Not a big difference perhaps, except for me, and I may do it without too many moral qualms, but the problem is that I may not be the only one to think in this way. If, for instance, there are fifty like me, then I and the other free riders will earn $50 from the rebate, and the others will each make $39. Note that I am still better off shirking than cooperating, because throwing in my $11 will not bring me anything more; it will instead cut my earnings to $40. In fact, whatever happens with the others, I am always better off shirking than cooperating. In other words, the only equilibrium in that situation consists of everyone shirking, so that everyone earns $0, thereby forgoing a possible gain of $89.

This seems extremely strange: here is a crowd of people who could earn $89 each, and who choose not to do so. But it does make sense: imagine for instance that there are ten thousand people in the group, but the cost of participation is raised to $1,000. This would increase potential benefits enormously, up to $9,000 per person, and yet it would make cooperation even more difficult. Indeed, just to recoup the cost of participation, one would need to find a thousand other people willing to bet $1,000 that they will not be the only ones; unless there is some bond between the participants, or some means of ensuring compliance, such people will be extremely hard to find. A general agreement on what is to be done is not enough: one needs the means to enforce the agreement. This may be called the principle of belling the cat: the advantage of having the cat belled is obvious, but no mouse is going to risk doing it unless compelled to do so. In fact, this may be the founding principle of modern states. According to the famous definition of Max Weber, the state is characterized by the fact that it has a monopoly on the legitimate use of violence. What should it be used for, except for coercing people into keeping their word in situations where the temptation to shirk would be too great. In the above example, if the ten thousand members of the group agree to be shot if they do not contribute $1,000 each, and if they appoint a person for the express purpose of carrying out the threat if they renege on their word (contributing perhaps $1 each for the policeman's pay), then they will have to comply, thereby earning $8,999 each, and the policeman will get $10,000 for doing nothing: his mere presence is enough. This is one reason why well-organized states need effective law-enforcement agencies.

The principle of belling the cat has wide applicability, especially in volunteer organizations, which have no means of enforcing compliance and rely exclusively on the goodwill of their members: after the initial enthusiasm, a few dedicated individuals end up doing all the work, while the vast majority barely bother to turn up at meetings. Trade unions face a similar problem, at different levels. To begin with, unionizing a firm or shop requires a majority vote; even if a majority of the workers feel that it would be to their benefit to unionize, it is quite another matter to step out, require that a vote be organized, and campaign for the issue, in the full knowledge that if it fails, you will suffer the backlash. Even if you are willing to take the risk, you are not sure that others will; the best strategy is then to wait for others to start the train, and to jump on when you feel there are enough persons on board. But this is a catch-22: if everyone waits for a clear majority to develop before joining, there can never be a majority. If, in addition, any advantage the union gets in collective bargaining is also available to nonunion workers, as is the case in France, then the benefits of joining are even less. If a union is ever established, it will have a hard time keeping up the membership; why pay the dues if one gets the benefits anyway? This may be why, in the United States, there are "closed shop" clauses, stating that only union workers may be hired, while in France unions try to provide their members with fringe benefits, unavailable to nonmembers, such as reduced rates for certain events or extended travel services.

Building an organization is a complicated story, and there are many more issues to worry about. We did say, for instance, that because of the principle of belling the cat, states need effective law-enforcement agencies. But what about collusion or corruption? In the last example we described, how are we to prevent one hundred of the ten thousand citizens to get out of paying their share by bribing the policeman, at a cost of $100 each? He gets away with $19,900, and each of them makes $9,900 (the $10,000 tax rebate, minus the cost of the bribe). We need policemen, but who will police the policemen? The answer may be an agency overseeing the police, which will raise the same problems on its own, or perhaps another police force, which will be kept at odds with the first one, so that each holds the other in check.

Humans are complicated beings. They are one branch in the tree of life on Earth, a particular result of the ongoing Darwinian process of descent with modification. They also see themselves as capable of intelligent behavior, whereby each individual seeks to further his own interests, while making informed predictions about the actions of others. There is no reason to believe that the dynamics of descent with modification or those of strategic behavior will by themselves lead human society to some kind of desirable outcome. In this chapter, we have reviewed much evidence to the contrary. In the struggle for life, or in the struggle for power, there is no reason why the survivors should be better than the dead, no reason why their victory would make the world better than it was. There is no invisible hand guiding these processes, dealing out victory to the most deserving. Chance is their true leader.

The End of Nature

NATURE IS INDIFFERENT. There is no one out there to watch over us. We are an animal species like so many others which have appeared and disappeared on Earth, and our Sun is a star like countless others in the universe. There is no hint in the laws of physics or biology of any special provision to take care of us. We are at the mercy of a cosmic catastrophe, perhaps a collision with one of the numerous celestial objects which gravitate near the orbit of the Earth, or of a biological one, such as a large-scale epidemic. Both have happened in the past, and will certainly happen again. Worse still, we are at the mercy of our own malice.

It is tempting to believe that, in spite of our individual failings, of which we are too aware, there will always be something to protect us from major catastrophes, as if some invisible hand would pull humanity back to safety at the last minute, as it teeters on the brink of extinction. As an extreme example of this way of thinking, some people claim that there is no need to worry about global warming, because God will not allow it. Closer to rationality, it is often claimed that a nuclear war is impossible simply because it is "unthinkable," that is, because the use of strategic nuclear weapons on a large scale would have such far-reaching consequences that the survival of human life, not to mention countless other living species, would be put at risk. Why would human beings willfully destroy their planet, like passengers sinking their ship in the middle of the ocean? Well, if that was the case, why did the United States and the Soviet Union spend enormous resources to prepare for a nuclear war over half a century? During all that time, and even now, thousands of intercontinental missiles, each carrying several warheads, have been held ready to fire at a moment's notice. No ingenuity has been spared to make sure that they reach their target: as I write, they stand ready in the depths of the ocean, on board flying aircraft, or in fortified bunkers. This is a tremendous organization, which can be activated within a few minutes, and if these weapons are intended never to be used, it would be a tremendous waste of resources.

The truth of the matter is that they were used (the two first atomic bombs ever built were dropped on Hiroshima and Nagasaki), they were always intended to be used (in case the other fellow used them first), and they came very close to being used again (during the Cuban missile crisis). Even now, in 2004, in the absence of any real challenge, the U.S. Energy Department has budgeted $6.5 billion for the nation's nuclear arsenal, 35 percent more in real terms than it spent in the years of the cold war, and on a par with the boom years of defense spending during the Reagan era. Not only will the old weapons be maintained, but new ones will be developed, "mini-nukes" or "bunker-busters," which will blur the distinction between conventional weapons and nuclear ones. It is not even clear that there is strategic thinking behind these developments: as usual, the technology is developed because it is available, and, by the same token, the weapons will be used because they are there— who knows by whom and in what circumstances?

History is replete with catastrophes caused by human action, and nuclear war would be just one among many. We have never seen any regulatory mechanism stepping in to prevent dangerous policies from developing and steering humankind back onto a safer course. It seems clear by now that Easter Island once was a lush place, and that it was turned into the barren place we know as a result of infighting: someone, over there, at some time, cut the last tree. Mesopotamia, the region between the Tigris and the Euphrates, was not always the desert we see today. In fact, it is the place where humans developed agriculture and invented writing, the cradle of our civilization. Either through overexploitation of the land, or through wanton destruction of the irrigation canals, Mesopotamia is now barren. Today, we are rushing toward similar catastrophes on a planetary scale. Global warming is the most striking example: even if we stopped producing any more carbon dioxide today, hundreds of years would elapse before its atmospheric concentration returned to its preindustrial level of 280 ppm. It is right now at 370 ppm, and it is expected to reach 745 ppm in 2100. By this time average temperatures will be higher than they are today, the estimated range being 37 to 41 degrees Fahrenheit, leading to some melting of the polar ice caps, so that the sea level will increase by an estimated half a foot to three feet, which is enough to wipe out some islands and countries like Bangladesh. There is always the hope that such dire predictions might turn out to be wrong, and this is an argument often put forward not to face the problem, but it should be kept in mind that there are two ways to be wrong: one can err on the good side as easily as on the bad side. In other words, the actual scenarios may turn out to be worse than expected (in fact, this seems to be happening), so that uncertainty actually strengthens the argument for doing something right now.

We no longer live in a "natural" world, but in an artificial one; we no longer adapt to our environment; we adapt our environment to us. The primeval forests are disappearing; so are the fish in the sea, and the ozone layer. Temperatures are rising, and there is no longer any place on Earth which does not carry some trace of human activity: even in the remotest places, one finds human-made pollutants, which make their way up the food chain. We are even becoming able to engineer our own species. We can seriously envision a time when it will be in our power to clone a given individual, or to choose the genes of our children, or to create chimeras, part human, part animal. These new possibilities strike at the root of all kinship relations, which have been from time immemorial the cement of our societies. Will we actually do these things, and what will the consequences be?

The possibilities which open up before us are not wholly without precedent, although they have to be sought in mythology as much as in history. Perfect doubles are not unheard of: nature produces twins once in a while, and stories of gods or demons taking on the identity of men, to enjoy their wives, for instance, are frequent. Choosing one's children has been done for a long time, by the simple device of discarding those one believed were imperfect; many primitive tribes, and the classical Greeks and Romans as well, killed newborns who exhibited some visible defects, and even in China today, baby boys suspiciously outnumber girls. Chimeras are revivals of centaurs, mermaids, sphinxes, harpies, and much of the bestiary of antiquity.

History and mythology hand down stern warnings against tampering with such things. The most famous one comes down from classical antiquity. It is the story of Oedipus, who becomes king of Thebes by killing the former king and marrying the queen. A great plague then befalls the city, and it is finally revealed, to him and the people, that he has killed his father and married his mother. Kinship relations have been tampered with; the tragedy then swiftly concentrates on Oedipus's children, who no longer know whether he is their father or their brother, and whether they are their own aunts or uncles. Collective and individual disaster follows. Oedipus blinds himself and becomes a vagrant, his mother/wife kills herself, his sons fight for the kingship and kill each other in battle, the city suffers a plague and a siege.

At a turning point early in the story, Oedipus meets a chimera, the Sphinx, a creature traditionally depicted with a woman's head and a lion's body. This is a dangerous encounter, since the Sphinx always ends up killing and eating his or her opponents, but this time Oedipus wins an apparent victory and kills the Sphinx. After this feat, he enters Thebes as a liberator, becomes king, and marries the widowed queen. The end of the story shows that it would actually have been much better for him to have died at that point, before the truth was revealed, and spared himself, his kin, and his city the disasters which he was to bring upon them. This is a clear warning that if one taboo is breached, namely the human/animal distinction, then others will follow, such as the mother/wife distinction, and the whole natural order will fall apart, with dire consequences for humanity.

Are such warnings valid for the modern world? We do not know. My point here is that we have to confront such choices, and there most probably is no turning back. We cannot pretend that the new possibilities offered by genetic engineering do not exist. There are many other technologies around, each of which has the potential to transform ourselves or our environment. Possible worlds are now crowding our doorstep. They are no longer purely virtual possibilities, which God envisioned and discarded in favor of the present (better) one. They are "clear and present dangers," or at least clear and present opportunities, which we can seize right now, and the effect of our actions will be felt for generations to come. For instance, one very real possibility is a warmer world, a planet where the environment has been profoundly altered by the greenhouse effect. The climate has changed, and so have dominant winds and currents; the northern ice cap has melted; the sea level has risen everywhere, swallowing islands and lowlands and bringing the coastline deep inland. We could also have continents without forests, seas without fish, savannas without game. We could, in fact, destroy our environment, simply by launching the thousands of missiles we maintain at great expense. The radioactive fallout will make vast regions uninhabitable, and the dust raised by the explosions will circle the Earth for many years, blotting out the light of the Sun and creating a nuclear winter in which many animal and plant species will perish. Again, such an event may be unthinkable, but it is not impossible: there are several examples in history of human societies destroying their environment, even though it meant destroying themselves. In the case of Easter Island, the destruction of the environment, notably the forests, through warfare and overexploitation took place over a very short period, probably two centuries, so that the people must have been well aware of what was going on. When the first Europeans arrived in 1722, they noticed that there were many wooden artifacts on the island, but no trees tall enough to make new ones. Someone, at some point, must have cut or burned the last tree, in full knowledge of what he or she was doing.

Which of these possibilities will become the new reality is our decision to make. This decision is urgent, for changes in the current environment are quickly becoming irreversible. In other words, we have to shape a new world, and to do it now. What a change since the time of Leibniz! In his view, the choice between all possible worlds had made once and for all, by God himself, at the time of Creation. Now this choice is ours to make. This is no longer a moral or theological problem; it is a matter of survival, not only for us, but for many other living species on Earth. The question is no longer one of the individual reaching peace of mind by recognizing God in his works and coming to terms with the presence of evil in the world. It is a question of human society today (or rather, a very small fraction of it) shaping the biological and social environment its descendants will have to live in for centuries to come. The first question is at best a moral one, and can be settled at leisure, while the second one is pressing. We are already experiencing the first effects of global warming, and by 2050, when the children born today will be middle-aged, it will be in full swing. And there is no one except ourselves we can turn to for the answer. According to the famous dictum, "If not us, who? And if not now, when?"

The lesson of Stoic philosophy was "to change one's thoughts rather than the order of nature." It may well be adapted to times when humanity was confronting natural forces immeasurably stronger than itself, but no longer now, when human activity is interfering with the geography and the climate and driving many species to extinction. One can doubt that there still is such a thing as the order of nature, at least at a scale that concerns humans: many of the phenomena that our ancestors would suffer blindly, as "acts of God," we can now control, influence, or predict. We can cure infectious diseases, direct ships out of the way of storms, build dams and dikes to prevent floods; we have cut forests and practically eliminated large animals from the face of the Earth, the few remaining specimens being closely monitored. There is very little now that could happen to us without our being able to do something about it. One could still envision humankind being destroyed by a natural event, but it would have to be at a planetary scale, such as the impact from a large asteroid. But even this is not quite a hopeless situation. Right now, the space around Earth is being monitored for large objects which could be on a collision orbit, in the hope that if such an event is predicted long enough in advance, some effort could be made to deflect it from its course, or to destroy it before the impact.

This active attitude toward nature is typical of humankind. Ever since the species *Homo sapiens* appeared on this planet, it has been making and improving tools and weapons, and it has been trying to harness the forces of nature to its own ends. The human is first and foremost an engineer, *Homo faber*, rather than a philosopher, *Homo sapiens*. Certainly, if an alien observer were looking at our planet, he would be struck by our technology rather than by our intellectual activity. Our African ancestors are known to us more by the tools and weapons they wielded than by the songs they sang and the myths they told. The technology gap that separates our modern nuclear plants from the primitive fire camps is tremendous, but we are speeding down the same path our forefathers started on. The pace has been accelerating throughout the journey, because we now have at our disposal resources in energy and scientific knowledge which our ancestors would not even have dreamed of, but we are still building tools and weapons, in the hope that they will help us live longer and better.

It has always been a mystery to historians why the development of science in classical antiquity did not trigger technological progress. Technologies remains pretty much the same all during that period, which covers roughly one thousand years. Science flourished during that period, at least during the first half, but this development does not seem to have spilled over to technology. A famous exception is Archimedes, who is reputed to have kept the Romans at bay during the siege of Syracuse by setting fire to their ships with mirrors and by devising all kinds of mysterious machines. Since he was killed by a Roman soldier when they finally stormed the city, that story was mostly understood as a lesson: scientists should worry about observing stars and not about winning wars. In other words, the scientist was seen as a philosopher, seeking knowledge for the sake of knowledge, far removed from the ways of the world, and with little social responsibility or impact.

The situation changes drastically at the time of the Renaissance. From then on, science went hand in hand with technology. Scientists prided themselves on being engineers, and engineers learned science to apply it. The measurement of time was the first example of a scientific discovery changing the technology. No conceivable improvement of the clepsydra or of the weights clock could lead to Harrison's chronometer, or to a modern quartz watch. A true discontinuity, a change of technology, was needed to exploit Galileo's theory of the pendulum. As we described it in the first chapter, Galileo's original theory turned out to be slightly wrong, and it took another technical improvement by Huygens, again built on theoretical ground, to build the first reasonably accurate pendulum clock, which set the stage for all subsequent developments. The notebooks of Leonardo da Vinci, that father of all engineers, are filled with drawings showing wonderful machines which would work for humans, transport them in the air or under the sea, feed them and protect them. Science was seen as a way of enhancing our powers, of putting more resources at our disposal, and thereby improving our lot. Scientists were no longer seen as mere stargazers: they were expected to contribute to the general welfare. Of course, I am oversimplifying the whole story. Galileo, for instance, drew more prestige from having discovered the five satellites of Jupiter and dedicated them to the Medici family than for his failed attempts at clock making. But it certainly is true that from that time on, scientists displayed an interest in concrete, even mundane, problems that they had not before. The great Pascal, for instance, built the first mechanical computer to assist accountants in making calculations.

Again, from a historical perspective, it is not clear what triggered the close association of science and technology that has prevailed ever since. One distinct possibility is that technological progress had already started on its own, and that scientists just boarded the train as it started moving. The Renaissance was also the time of the Italian wars, when French and Spanish armies fought for dominion over the peninsula, and, in the beginning at least, the technological superiority of the French in guns and artillery was so overwhelming that their opponents had to adjust very quickly. This created a great interest for the study of ballistics, and a very favorable context for Galileo's work on falling bodies. In the same vein, field glasses were first developed for military use before Galileo had the idea of turning them to the night sky. Whichever came first, technology or science, the end result is undisputed: modern science gets some of its inspiration from technological problems, and technology benefits from scientific discoveries. This alliance is well displayed in Diderot's *Encyclopédie*, which records in minute detail the science and technology of the late eighteenth century. It was a fully coherent body of knowledge, and all who prided themselves as cultured were supposed to be aware of it, not only scientists and engineers, but also gentlemen and "femmes du monde."

When Euler, Maupertuis, and Lagrange invented and developed the calculus of variations, their aim was not only to lay solid mathematical foundations for classical mechanics, but also to devise the best possible solutions to a series of technological problems. The mathematical methods they devised are now used to find curves, or more generally shapes, for which a prescribed criterion is stationary. If we take as a criterion the least action principle, thereby putting ourselves in God's shoes, as Maupertuis believed, the calculus of variations will enable us to recover all the laws of classical mechanics. But the same techniques apply to engineering, and to any situation where the aim is to build machines or devise processes which have to function in the most efficient way possible. This will be done by defining a suitable criterion for performance (the higher the numerical value of the criterion, the better the performance), and to seek the machine or the process that will maximize that criterion.

The moment when the scientists became engineers was a historical turning point. Knowledge was longer sought to understand God's ways and the wonders of his works. It was used to build machines that could assist human beings in their works and endeavors. The difficulties that beset us in earlier chapters disappeared. The least action principle had appeared shrouded in mystery, and we had been left to wonder about its deeper significance. But when the engineer or the designer picked his own criterion, so as best to represent the technological problem at hand, away went the metaphysical discussions, and the subtle distinctions between stationary points and maxima. No longer interested in stationary points, engineers wanted solutions which truly maximized the performance. Alternatively, if we think in terms of cost rather than performance (which is just the flip side of the coin), they were looking for cost-minimizing solutions. To convey the idea of minimizing or maximizing, and that they were not interested in stationary points, they said that they were *optimizing*.

Fermat's proof of the law of refraction really consists in solving an optimization problem, namely to find the quickest path between two given points. Maupertuis considered the whole world as an optimization problem, but, as we have seen, he was wrong. The first scientist to solve a technological problem by optimization was, again, Newton. In his *Principia*, after establishing his celebrated results on the inverse square law, he turns to a much more mundane question: what is the best shape for a bullet? what shape should we give an object in order to minimize air resistance? Newton begins by giving a mathematical expression for air resistance, no small feat at the time. He then restricts attention to objects which have a symmetry of revolution, that is, the profile of which is not altered by rotating around an axis. The shape of such an object is entirely defined by its profile. Newton then finds, for any prescribed height and surface area, the symmetrical object which minimizes air resistance. In other words, given the length and bore of the bullet, he finds the most efficient symmetrical object.

The shapes Newton found were unexpected: the tips of his bullets were flat. One would have expected a sharp tip, in the idea that any frontal area would slow down the bullet. The explanation lies in the particular expression Newton gave for air resistance. He derived it by considering air as a multitude of independent particles, each of which slows down the solid when it hits it: air resistance to the motion is nothing but the resulting sum of these elastic shocks. By so doing, Newton neglected the interaction between particles and the fact that shocks are not elastic, so that his formula is valid only for slow speeds, and strangely enough, for very fast speeds (several times the speed of sound). In other words, it is better suited for designing spacecraft than for designing bullets.

One can only marvel once more at Newton's genius. There he was, finding an optimal shape, at a time when Euler and Lagrange, the founders of the calculus of variations, were not even born. To put his

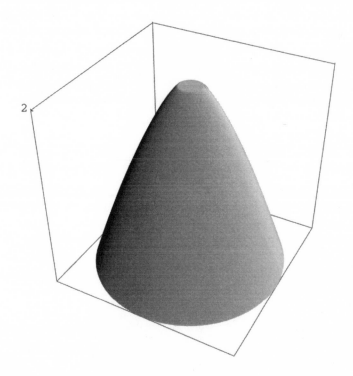

20. NEWTON'S PROBLEM OF MINIMAL AIR RESISTANCE Newton asked what shape a bullet (or any object moving through air) should have to minimize air resistance. This is the solution he found in the case that the bullet was supposed to be as wide as it was long. If the bullet is allowed to be longer, the flattened part at the tip disappears. If it is supposed to be shorter (as is the case for a space shuttle), then the flattened part widens.

achievement in perspective, let me point out that today, more than three centuries later, we still do not know which is the shape which will minimize air resistance (as given by Newton's mathematical formula). What Newton found was the optimal profile when it is assumed that the solid is symmetrical around an axis, but it has been proved recently that there are nonsymmetrical solids which offer less air resistance, although it is not known which is the optimal one.

Ten years after the *Principia*, we encounter the brachistochrone problem, which we have described earlier, and which engaged Jacob and Johann Bernoulli, the two brothers from Basel, in a long quarrel about priority. Let us recall that the problem consisted in finding the profile that will bring a sliding body down from a given height in the shortest possible time. Clearly, this had no more than academic interest, and it is difficult to understand why the glory of having solved it would lead two brothers to feud, unless one realizes that it was really about conquering new territory, much like an explorer who is the first one to land on a sandy beach unfurls a flag and claims the whole continent for his king. Over the next century, the methods developed to solve particular problems would be systematized and consolidated by Euler and Lagrange into a new branch of mathematics, called the calculus of variations. At the end of the eighteenth century, the general equations were known, together with a wealth of examples, and scientists were familiar with the idea that there were mathematical tools which could help them find curves (and, more generally, shapes) which would maximize a given criterion.

According to the spirit and knowledge of the times, solving a problem in the calculus of variations meant writing down the Euler-Lagrange equations in the particular case at hand, and then finding their solutions. As we noted in earlier chapters, the latter can be done only in the case of integrable systems, which are a very small fraction of all possible systems. The founding fathers of the calculus of variations were not aware of this situation, and much time was spent identifying the very few problems which could be solved in that fashion (as opposed to the very many chat could not) before the underlying mathematical problems were tackled, toward the beginning of the twentieth century.

Here there appeared an important difference from classical mechanics, in which the emphasis lies on the equations of motion, for which a general solution is sought. In the calculus of variations, however, one is not interested in finding all the solutions of these equations, but only the particular one (or ones) that will satisfy additional conditions. For instance, in classical mechanics, one is interested in finding the motion of a point that is not subject to any force at all. It turns out that such a point would move with constant velocity, along some straight line. On the other hand, in Euclidian geometry, one is interested in finding the shortest path between two given points. This path lies on a very particular straight line, the one joining A and B, and is in fact the segment between them. My point here is that one may conceivably solve the second problem without solving the first one, even though, in this deliberately simple example, both can be solved so easily that it is difficult to distinguish them. In other words, there is a difference between finding all possible trajectories of the motion, as classical mechanics requires, and finding the unique trajectory starting at a given point and ending at another.

In classical mechanics, one is interested in solving the equations of motion, that is, in finding all possible trajectories of the system for all possible initial conditions. If these equations are not integrable, this does not mean that the corresponding trajectories do not exist: it simply means that we do not know how to compute them efficiently. From this point of view, as Lagrange pointed out, the least action principle is not central to classical mechanics. It is a just a concise way of writing down the equations of motion, and of finding out whether the system is integrable or not. In the calculus of variations, on the other hand, it is the criterion which is central: the whole point of the problem is to maximize it (or to minimize it, if it is a cost). This means that the solution must satisfy not only the Euler-Lagrange equations, but also some boundary conditions, such as joining two given points, which will distinguish it among all the other solutions of the Euler-Lagrange equations. It is not even clear that such a solution would exist: some kind of compatibility is required between the equation and the boundary conditions. These difficulties, which were glossed over at the beginning of the theory, became more and more evident and embarrassing. In 1900, David Hilbert listed the most important unsolved problems in mathematics at that time. That list of twenty-three problems has been extremely influential in the development of mathematics in the twentieth century, and among them we find the question "whether the problems in the calculus of variations have solutions or not, the notion of solution being interpreted in a broad enough sense."

Most of Hilbert's problems have been solved by now, including this one. Thanks to the work of the Italian Leonida Tonelli (1885–1946), and of the Frenchman Henri Lebesgue (1875–1941), we have an almost complete theory of the calculus of variations, which enables us to say whether a given problem does, or does not, have a solution. In the second half of the century, powerful numerical methods have been devised to compute such solutions, when they exist. There remain, however, some unsolved problems. Classical mechanics dealt with rigid bodies; this is unrealistic, for in practice a solid body will deform when forces are applied to its surface, and this deformation will generate stress in its interior, possibly leading to rupture. Very little is known, on the other hand, when we are dealing with deformable bodies. This is the realm of continuum mechanics, which can be formulated as a problem in the calculus of variations, but for which there is as yet no satisfactory theory.

Looking away from these difficulties, it is fair to say that the calculus of variations has firmly established optimization as a central concept in modern mathematics. To optimize is to find, among all possible solutions to a given problem, the one which will maximize the performance, as defined by a suitable criterion. The first historical example is probably Newton's shape of least air resistance. Since then, engineers have learned to build structures, such as bridges, boats, buildings, or airplanes, at minimal cost for prescribed performance, or at maximal performance at prescribed cost. Nowadays, we have a continuous stream of design problems resulting from technological advances; we have the theory which enables us to phrase them as optimization problems; and we have the computers which will help us find the solution. The scope of optimization theory has also extended far beyond engineering, into economics, management, and finance.

The turning point probably occurred during WWII, when the logistics of producing and distributing equipment, ammunition, and food for millions of soldiers scattered over the planet strained the management capacities of the human brain. The idea of formulating these problems in mathematical terms, and of using optimization theory to solve them, then took hold, leading to the birth of a new field of knowledge, called operations research. Fifty years later, the continuing progress in numerical methods and in computer technology has put solutions to many optimization problems within easy reach of engineers and managers.

As a typical example, consider the problem an airline faces when rotating planes and crews. For each flight, a plane and a crew must be made available at a prescribed time and place. There are numerous constraints to be satisfied: no pilot can fly more than so many consecutive hours, or fly more than so many hours a month, or stay away from home more than so many days on end. Planes must be maintained after so many hours in flight and undergo a complete overhaul every year. Even so, there are occasional failures, and backup crews and planes must be available as soon as possible. The problem is to find a schedule which will accommodate all these constraints, at the lowest possible cost in terms of fleet and labor, typically an optimization problem. As a second example, consider the problem of sending a spacecraft to Mars. This is not a question of steering directly to the goal, with the gas pedal down until it is reached: there is not enough fuel on board for that. The reactors are switched on during the launching period, and back again toward arrival, in order not to crash on landing. For the remainder of the flight, they are turned off, and it is inertia, compounded with gravitation, which propels the spacecraft. If it veers off course, the reactors are switched back on to put it back on track, but these should be short episodes. Of course, Mars is moving while the spacecraft is on its journey, which may last quite a long time. Hence a beautiful optimization problem: what is the trajectory to follow (or, in what direction is the spacecraft to be steered away from the Earth) in order for the trip to Mars to require the least possible fuel? This is the criterion of least consumption. Others are possible, such as the least time criterion: given a certain amount of fuel, which is the trajectory to follow in order to reach Mars as soon as possible?

Note the difficulty here, which is central to the calculus of variations: whatever the criterion, be it least consumption or least time, we will know its actual value only at the end of the journey, when we land on Mars. As long as we are on our way, we do not know that final value which is to be optimized, although we have some idea, which becomes more accurate as the destination gets closer. To steer the spacecraft efficiently, we need some instructions about what to do right now, not what the overall trajectory should look like when we have reached the end. This is precisely what the calculus of variation will do for us, at least in its modern form, discovered by the Soviet mathematician Lev Pontryagin (1908–1988), and formulated as a principle which bears his name. Applying Pontryagin's principle tells us what to do at every moment of the journey, whether to turn on the engines, and if so in which direction to steer.

As a final glimpse into present-day optimization theory, let me mention problems with inherent uncertainty. It is not always the case that the state of the system is known with absolute certainty: measurements are never wholly accurate, and to position a spacecraft, for instance, one must take into account several observations, all affected by background noise. This is particularly important during takeoff, when there is great instability, and any slight deviation has to be spotted at once; otherwise it amplifies and becomes unmanageable.

Steering optimally with noisy observations is called filtering, and it is an essential part of aircraft and space engineering. There are also many situations where there is uncertainty about the payoff. In finance, for instance, no one knows what a portfolio that is worth $1,000 today will be worth one year from now, and yet there are many investors out there, professionals and amateurs, trying to make money out of the market: they are trying to solve an optimization problem with uncertain payoff. To solve these problems, the classical methods of the calculus of variations combine with the more recent tools of probability theory, resulting in a very active field of applied mathematics, called stochastic control.

Looking back at the sophistication which optimization techniques have reached today, one wonders whether they could be used to solve not only industrial and managerial problems, but some economic and social ones. Certainly, the way society distributes wealth and power is at least as important for our everyday life as the way it organizes industrial production and distributes consumer goods. Could our governments use the concepts and methods of optimization theory to assist them in their task, as engineers do? Could one try to organize society efficiently, as one organizes an industrial system?

Of course, trying to construct a theory of society, let alone to apply it, is much more ambitious than trying to construct a theory of the universe. But mathematical modeling has been so successful in the natural sciences that one may hope for some measure of success as well in applying the same method to the social sciences. Before we do that, there is a question to be answered: is it possible to alter the organization of society, or to build a new one, as we repair a broken-down machine or build a new one?

This is not obvious. Who will do it, why, and how? In certain societies, living very close to the edge of survival, and under very strong constraints from their environment, there is practically no room for change. I am thinking, for instance, of the Inuits, who have to survive the long Arctic night, or the Alakalufs and Oonas, who spent their lives on canoes roaming the inhospitable waters of the Magellan Strait, until they were hounded down and exterminated. In richer societies, where subsistence is not such a pressing problem, there are other constraints, of a social and psychological nature, which have become a fact of life to their members. They are born and bred into a certain social organization, and they would no more think of changing it than changing the color of the sky. In France, for instance, at the time of the monarchy, the king was seen to be God's own representative on Earth, as his father and his ancestors were before him. Religion supported that view, as it supported the Inca in Peru, the Son of Heaven in China, the sultan in Turkey. But then came the French Revolution. It showed that the social order can indeed be changed, and the hopes it raised have been with us ever since. In the words of the philosopher Richard Rorty, "About two hundred years ago, the idea that truth was made rather than found began to take hold of the imagination of Europe. The French Revolution has shown that the whole vocabulary of social relations, and the whole spectrum of social institutions, could be replaced almost overnight. The precedent made utopian politics the rule rather than the exception among intellectuals. Utopian politics sets aside questions about both the will of God and the nature of man and dreams of creating a hitherto unknown form of society."[1]

One might argue that the actors of the French Revolution were far from having such a radical position as the one we describe. They were believers, not in the Christian god, but in some kind of natural order, and they thought they were merely restoring the natural order of society. But there have been too many revolutions since for us to entertain any more the belief that there is a natural order of society. The Constitution of the United States, and the several constitutions of the former Soviet Union, have regulated the lives of millions of individuals according to widely different principles. They are a testimony that social organizations are malleable, and that political systems are provisional. They do not reflect some celestial or natural organization; they are products of history, like works of art and scientific knowledge.

The fact that political systems are human creations had been known in classical antiquity and had been forgotten, until the Italian Renaissance brought it back to light. The Greeks invented the polis, the city-state ruled by its own citizens rather than a ruler. This was a fundamental innovation, which would bear fruit two thousand years later in Europe. This is how S. E. Finer, in his great *History of Government from the Earliest Times*, puts it: "From the beginning of recorded history in Sumeria and Egypt—for some two-and-a-half thousands of years—every constituted state had been a monarchy: not only in the known world of the Middle East and Eastern Mediterranean, but in the worlds of India and distant China too. These monarchs had all been absolute, and godlike too, except for the Jewish kingdom where God ruled the kings. Suddenly there was government without kings of

1. *Contingency, Irony, and Solidarity* (Cambridge: Cambridge University Press, 1989), 3.

god. Instead there were man-made, custom-built republics of citizens."[2] And later on, he explains that the Greeks "addressed the form of government directly. More, they were completely self-conscious of this. As a corollary, their polities became completely contrived instruments to achieve consciously expressed goals, and indeed were often deliberately reshaped. This, in short, is the very beginning of 'the state as a work of art.'"

The Italian city-states of the late Middle Ages exhibit the same creativity. Unfortunately, the Greek and the Italian experiments were not allowed to continue: both were cut short by foreign intervention, Philip of Macedonia in Greece, and Charles V of Hapsburg in Italy. Military conquest and foreign hegemony spelled the end of the city-states, and of the extraordinary period of creativity which they fostered in all fields of human endeavor, the age of Pericles in Athens, and the time of the Renaissance in Italy. We are lucky enough that the two great historians whom we introduced in the previous chapter, Thucydides for Greece and Guicciardini for Italy, recorded the troubled times that they went through, and transmitted the main results of these two great experiments.

Both men depict terrible events, wars dragging on for a lifetime, destroying cities and laying waste the countryside, leaving in their wake the wreckage of thousands of lives. In the words of Thucydides, "The greatest event of former times was the Persian war, and yet this was quickly decided in two sea-fights and two land-battles. But the Peloponnesian war was protracted to a great length, and in the course of it disasters befell Hellas the like of which had never occurred in any equal space of time. Never had so many cities been taken and left desolate, some by the Barbarians, and other by Hellenes themselves warring against one another; while several, after their capture, underwent a change of inhabitants. Never had so many human beings been exiled, or so much human blood been shed, whether in the course of war itself or as the result of civil dissensions."[3]

Guicciardini's *History of Italy* opens in similar fashion: "I have decided to write down the events that happened in Italy in our times, after the French arms, called into the country by our own rulers, began to throw it into extreme turmoil and agitation. Because of their variety and importance, these were events which are well worth being recorded, and which had terrible consequences. During all these years Italy has suffered all the

2. S. E. Finer, *The History of Government from the Earliest Times* (Oxford: Oxford University Press, 1997), 316.

3. *History of the Peloponnesian War*, 1.23.

calamities to which poor mortals are subject, sometimes because of God's righteous anger, and sometimes because of the impiety and evil of their fellow men."[4]

As we saw above, both Thucydides and Guicciardini held high military commands, and their ultimate defeat hit them especially hard. Thucydides commanded the Athenian army in the North, in Thrace, and was sent into exile in 424 for letting the city of Amphipolis fall to the enemy. Very little is known about his life after that. In 1527, Guicciardini was in charge of the pontifical army, allied with the Venetians and the French against the Spaniards. This alliance was to a large extent his doing, and should have been strong enough to drive the Spaniards out of Italy. But dithering on one side and boldness on the other led to the opposite outcome: Rome fell on May 5, 1527, and the pope became a prisoner in his own palace. Guicciardini's dream of a free Italy was over, and he retired from public life a few years later to start working on his *History*.

Why do men who have held such high office start writing after their public careers have ended in failure? To show that such catastrophes could have been avoided, that their causes are not to be sought in some divine will but in human folly, and in the hope that future generations will learn from the mistakes of their predecessors. In a famous sentence, Thucydides says, "Whoever shall wish to have a clear view both of the events which have happened and or those which will some day, because of human nature, happen again in a similar way—for these to deem my history profitable will be enough for me."[5] As for Guicciardini, after the opening we already quoted, he goes on as follows: "From the knowledge of these events, so full of diversity and import, every one will be able to derive many instructive lessons. They will show by innumerable examples how unstable human affairs are, like a sea churned up by the wind."[6]

In other words, neither of them is willing to give up. They have seen the ways of the world, they have participated in decision-making at the highest level, and they have seen how soldiers behave in wartime. The world they have seen is so bad that one can only want to change it for the better, be it ever so little. But is it possible? This is really what their work is about. Both their histories show how ill-conceived decisions, made without proper consideration or under the pressure of passion, have led to disasters. They also show how a few great individuals, Pericles in Athens, Lorenzo de' Medici in Florence, have managed, through patient and intelligent efforts, to secure for their people long periods of peace and

4. *History of Italy* (1537–1540), 1.1.
5. *History of the Peloponnesian War*, 1.22.
6. *History of Italy*, 1.1.

prosperity. Their work was later undone by incompetent or imprudent successors, and this is the story that Thucydides and Guicciardini are telling. The lesson is that history is not running blindly along, that individuals have the power to veer its course one way or another. The social world is not driven by natural laws and randomness alone, as the physical world is, but also by human wills. We are actors in history. The fate of humankind does not lie in the hands of God, but in our own.

At the time of the Renaissance, expressing such an idea would have been as dangerous as claiming that the Earth moves around the Sun. So Guicciardini goes to great lengths to disguise it as a pious thought, as just another of the wonderful ways God finds to have his will done in the world. In the first of his *Ricordi*, he strikes a very delicate balance between faith and reason. Although he tries to give the impression that God is active in human affairs, he leaves humans firmly in charge:

> What the believers say, that whoever has faith will accomplish great things, and, as the Gospel says, that whoever has faith can command mountains, happens because faith creates obstinacy. To have faith is nothing else than believing strongly, with quasi-certainty, things which either are not reasonable, or if they are, believing them with a stronger conviction than reason would allow. In that way, whoever has faith becomes firm in whatever he believes, and follows his way with an intrepid and resolute stride, disdaining difficulties and dangers, and ready to endure every extremity. Since the turn of events in this world depends on a thousand chances and accidents, it may well happen that, in the course of time, some unexpected help will come to whoever has persevered in his obstinacy, the source of which is faith.[7]

Thucydides and Guicciardini teach us that there is no hidden meaning to be found in history, that the course of events is not predetermined, and that the conscious actions of individuals can steer that course in new directions. The consequences are important. On the one hand, every treaty, every constitution, every institution, every state, is provisional. They will flower and die, for they are not reflections of some divine or natural order, but agreements between human beings, who are subject to death and the multiple contingencies of life. Sooner or later, every empire will perish. Thucydides witnessed the end of the Athenian empire and Guicciardini the end of the Florentine Republic. We have seen the end of the European colonial empires, and the fall of the Berlin wall. Human affairs are fluid and in constant flux; powers that look formidable one

7. *Ricordi*, 1.

day can vanish the next morning, like ghosts fading away when the cock crows. This permanent ebb and flow keeps bringing up opportunities which talented individuals like Pericles or Lorenzo the Magnificent then use to steer the course of history. In Shakespeare's words:

> There is a tide in the affairs of men
> Which, taken at the flood, leads on to fortune;
> Omitted, all the voyage of their life
> Is bound in shallows and miseries.
> On such a full sea we are now afloat
> And we must take the current when it serves.
> Or lose our ventures. [8]

These ideas will be carried further by the Machiavelli in Italy, Montaigne and Pascal in France, and Gracian in Spain. Montaigne's *Essays*, which went through several editions during his lifetime, and Pascal's *Thoughts*, which were collected after his death from notes for an unfinished book, tend to show that social life is regulated by conventions. They record the vast diversity of customs and rules across peoples and times, and the enormous disparities in what is considered normal behavior across societies, with the clear implication that the source of social regulations is not to be traced to some divine source or some permanent feature of human nature. There is nothing universal or permanent to be found there: nothing is so outlandish or repugnant that human beings have refrained from doing it to each other. Social regulations and institutions are mere human artifacts, with no other justification for their existence than the fact that they make social life possible by enabling its members to adjust their behavior to that of others. The fact that they are around, that everyone knows them, and that everyone knows that everyone else knows them, creates common expectations and enables us to anticipate how others will react when we interact with them.

Pascal is full of examples to illustrate this. Here is one: "The greatest of all ills is civil wars. They are certain, if merit is to be rewarded, for all will claim that they are deserving. The ill to be feared from a fool, who inherits a position by birthright, is neither as great, nor as certain."[9] Here is another: "Whatever in the world is most unreasonable becomes most reasonable because of the unruliness of men. What could be less sensible than choosing the first son of a queen to rule a state? To pilot a ship, no one would choose the highest born passenger: such a law

8. *Julius Caesar*, act 3, sc. 2
9. *Thoughts* (1670), frag. 295.

would be unjust and ridiculous. But because they are unruly, and will so remain forever, that law becomes reasonable and just; for who is to be chosen? The most virtuous and the most expert? We shall immediately start fighting over it: everyone claims to be that most virtuous and most expert person. Let us therefore attach that power to something that is not controversial. He is the elder son of the king; the fact is clear, there is no arguing over it. Reason cannot do any better, for civil war is the greatest of all evils."[10]

Power is no longer seen as inheriting its legitimacy from some divine authority; it is a mere convention which we adhere to because we are born and educated into it, and because we see that others conform to it. Its strength lies in the fact that we believe that others believe in it: power is no more than the illusion of power. The exercise of power is a constant fight to keep up appearances. A related thought is expressed by Balthazar Gracian in Spain: "Things are not taken for what they are, but for what they appear to be. Very few people look into the inside, and almost everyone is content with appearances. It is not enough to have meant well, if the action looks bad."[11] and by Niccolò Machiavelli in Italy: "It is not necessary for a ruler to have all the qualities I have described, but he must seem to have them. I would even say that if he really had them, and kept observing them, they would be to his disadvantage, whereas if he just pretends to have them, they are to his benefit. You should seem to be compassionate, faithful, human, generous, honest and pious; you could even be so, provided you are resolved, in case you need not to be, that you can and will do the opposite."[12]

So appearances come before substance. But what is substance if all that others can perceive are appearances? Does there really exist, beyond visible reality, some invisible soul or conscience, which the individual carries around? How helpful is that assumption? Do we need it, or can we do without, like Laplace, who famously answered, when the emperor Napoleon asked him what place was left for God in cosmology, "Your Majesty, I did not need that assumption"? Is it possible to study human beings from the outside, relying only on what we actually see, that is, their behavior? This is another Copernican revolution in the making: the Earth is no longer at the center of the physical world, and God will no longer be at the center of the social world. Thucydides, Guicciardini, and their successors teach us that history is as chaotic as any physical system, and that there is no God-given or natural order.

10. Ibid., frag. 296.
11. *Oraculo manual y arte de la prudencia* (1647), frag. 99.
12. *The Prince* (1515), in *Oeuvres complètes* (Paris: Gallimard), chap. 20.

This immediately raises the question: what are we to put in its place? History shows that priests and laypeople, civilians and soldiers, peasants and city dwellers all follow their own perceived interests, each reacting to the actions of others and trying to anticipate them. Which institutions will enable society to function in the most efficient way? An important part of the work of Thucydides and Guicciardini is devoted to comparing social regimes and constitutions, aristocratic Sparta and democratic Athens for the former, not to mention the Persian Empire with its semidivine King of Kings and the many reorganizations that the Florentine Republic underwent for the latter. Their concern was to find the best possible constitution. This line of thought gave birth to the great constitutions of the eighteenth century, the United States and the first French Republic.

The Renaissance raised two related questions. The first one, which we have investigated at length, is whether the natural world can in any way be considered as the best possible. We are now facing the second one: what is the best way to organize society?

(CHAPTER 9) The Common Good

THE FUNDAMENTAL DIFFERENCE between the physical and the social worlds is purpose. When people do something, they have a purpose in mind; at least, if they are asked why they are acting in that particular way, they are able to supply a reason. Human actions are directed toward an end; physical events are not. If fire breaks out in a room, people and smoke will leave it. The dynamics are similar, but the reasons are quite different. People get out because they want to get out, whereas hot air merely conforms to the laws of physics. By so doing, it may well be that it maximizes some quantities, such as entropy, or minimizes others, like action, but it is not intentional, like someone rushing out to escape the burning room. Since Galileo, we have had a theory of physical events, based on the premise that they conform rigidly to certain mathematical laws. Developing a theory of human actions, based on the premise that they all have a purpose, took much longer. It was one of the main scientific achievements of the twentieth century, in close connection with the progress of economic theory. Its history, which basically moves from Central Europe to the United States after the Nazi takeover, is too long to retrace here; let me just mention the name of John von Neumann, who was a central figure in that field of science as in so many others. His 1944 book with Oskar Morgenstern, *Theory of Games and Economic Behavior*, has been tremendously influential in shaping the future of the social sciences, particularly economics. Let me now try to describe what the theory looks like today.

In line with the thinking of Thucydides, Guicciardini, and their successors, the theory does not try to promote some idea of a universal Good, valid across societies and times. There is no absolute judge of what is good, or bad; only the individual can tell what is good, or bad, for him or her, and the best the theory can do is to record his or her preferences. The basic theoretical premise of the economic approach to human behavior is that individuals have linear preferences: each of us is supposed to be able to classify, by order of preference, all possible events, starting from the most preferred one and going down. It is a complete list of all possibilities, and if event B is farther down my list than event A, it means that I would rather have event A occur than event B. Typically, different individuals will have different lists; if events A and B are on both our lists, it may well be that they occur in different order, because I prefer A to B and you prefer B to A.

Individual preferences are simply a matter of record. No set of preferences will be disallowed because they are unreasonable or immoral; if I prefer peanut butter to foie gras, and dog meat to peanut butter, so be it. No claim is made that one list is better than the other. There are, however, some problems with this definition. Basically, it consists of confronting each individual with a list of hypothetical situations and asking him or her to rank them. Well, it might be difficult to project oneself into situations which are quite real, but about which I know very little (what would my life be like if I were a movie star?), and even more so into situations about which no one has any experience at all (twenty years ago, no one except James Bond used a mobile phone). On the other hand, people make decisions based on what they know now, and not what they will learn later. Defining individual preferences simply as a record of present tastes has its limitations, like all models, but it is realistic enough to serve its purpose.

It seems that, since each individual is the only judge of what is best for himself, the best of all possible worlds would be the one that every one would prefer. That would be a world where every individual's wishes would be realized, which is the way most people would define paradise. Unfortunately, this is not a possible world: we cannot all be movie stars or great artists or successful businesspeople. Most human beings on this planet would be satisfied to be shielded from hunger, sickness, and war. Things were different in the Garden of Eden, although the problem there clearly was boredom. Curiously enough, the French philosophers of the Enlightenment, who fought so strongly against the church, recreated the Christian myth of paradise lost by imagining that we originally lived in peace, from the bounty of nature, until civilization corrupted us and brought all the evils that we know. In the words of Jean-Jacques Rousseau, "I see [primitive man] eating under an oak, drinking from a stream, making his bed under the same tree that provided his meal, and all his needs are satisfied." Well, it has been a long time since manna last fell from heaven. We cannot live alone; we rely on others to produce the stuff of our material and intellectual life, and we have to organize society so that its members will cooperate toward the common good.

But what is the common good? Individual preferences may very well be in opposition. Rousseau thinks that social life emerges in the idyllic state of nature because individual preferences turn out to be in unison: while no single hunter is strong enough to track down a deer by himself, several can cooperate to accomplish this feat. So far, there is a common interest to be pursued. But once the deer is killed, the problem of sharing it appears, and individual interests diverge: no one can get a bigger share except at the expense of someone else. There is no longer a clear common interest, and there is no obvious sharing rule. Sharing equally among the hunters, for instance, may not be the best thing to do, either because it is not feasible (there are different morsels to the deer, and different tastes among hunters), or because it is not fair (whoever was the first one to see the tracks probably deserves some reward), or because it does not meet the collective needs (are bachelors to get the same share as those who have families to sustain? what about the children, women, and elderly who cannot participate in the hunt?).

Many philosophers have assumed the problem away by postulating an identity between individual aspirations and the common good, suitably defined. Rousseau, and the French revolutionaries after him, thought that democracy was the solution to that problem. In the introduction to his *Discourse on the Origin and Foundations of Inequality among Men*, he wrote, "I wish I was born in a country where the sovereign and the people could have only one and the same interest, so that all the movements of the state would always be directed to the common good; since this cannot happen unless the sovereign and the people are one and the same person, this implies that I wish I was born under a democratic government, suitably tempered." Unfortunately, things are not as simple as that. Even in democracies, there are many divergences of interest, and majority rule does not solve that problem. The French Revolution quickly turned into a civil war against the supporters of the old order, and in today's large democracies, we find that a majority of citizens, mostly the poor and dispossessed, have given up on voting, so that their interests are not taken into account. At least they could vote if they so chose, so that there is some measure of popular control on government; according to Winston Churchill (himself an aristocrat) "Democracy is the worst of all régimes, except for all the others." It is by no means perfect, and indeed I think that improving it should be one of the main concerns of our times. We have seen democratic leaders go to war against the wishes of their people, and there is no institutional way in which the poor on this planet can make themselves heard.

The British economist Francis Hutcheson, in the eighteenth century, coined a formula that was to be much quoted afterward. He defined the common good as "the greatest welfare for the greatest number." This certainly is a brilliant formula, but it is useless. How helpful would it be to me, if I have to share a cake, to be instructed to give "the greatest share to the greatest number"? Either I give the greatest share, that is the whole cake, to one person, or I divide the cake among all, but then everyone gets a small piece: I do one or the other, I cannot do both. Of course, increasing welfare is not exactly the same as sharing a cake. If I improve the quality of air, for instance, everyone will benefit, whereas if I give a piece of cake to only one person, no one else will enjoy it. So there is a contradiction in Hutcheson's definition, and it cannot serve as a definition of the common good or the public interest. How then are we to define it?

The common good is quite an elusive concept, and in fact, in any situation, different people have different opinions on what society should be doing. So let us just give up on trying to define the common good, or the public interest, and throw the decision back to the citizens. The French Declaration of Rights stated that "law is the expression of the general will." But a nation, unlike an individual, does not have an organ to mouth its will, and if we want to find out what it is, if anything, we must put in place institutions and procedures directed to that purpose. Is there an optimal way to do so?

Well, by now humankind has had extensive experience with government, and many procedures have been developed for collective decision-making. In 1785, however, the marquis de Condorcet showed that some of the most popular procedures, like majority voting, could lead to the assembly contradicting itself. His 1785 book, *A Discussion on the Application of Analysis to the Probability of Decisions Taken by Majority Voting*, is a milestone, for it is the first time mathematics was used to model human behavior. It contains the first example of what is known today as Condorcet's paradox. Suppose there are three candidates for office, Andrew, Brian, and Catherine. One-third of the voters rank them A, B, C; one-third B, C, A; and one-third C, A, B. There is then a two-thirds majority preferring A to B, and (a different) two-thirds majority preferring B to C. Two motions could then be passed, stating, "This assembly prefers Andrew to Brian," and "This assembly prefers Brian to Catherine." It would then be expected that a third motion, "This assembly prefers Andrew to Catherine," would pass as a logical consequence of the others, but in fact it would not. It would be rejected by a two-thirds majority.

It turns out that this is a problem with majority voting, and that other procedures are immune to Condorcet's paradox. Such a procedure was devised by the chevalier de Borda, who proposed it in 1785 to elect members of the French Academy of Sciences. It was adopted and used until Napoleon (himself a member), who thought it too democratic, used his authority as emperor to change it. In Borda's procedure all voters rank their choices by order of preference, from their first choice to the last. One then sums, for each candidate, the ranks he has been given by the voters. Suppose, for instance, there are three candidates, A, B, and C. If candidate A has been ranked first 13 times, second 18 times, and third 4 times, his score is $13 + 36 + 12 = 61$. The candidate with the lowest score wins the election.

There are great advantages to Borda's procedure. On the one hand, it enables voters to give some sense of the intensity of their preferences: if there are two candidates only, this effect is not felt, but if there are seventeen, the one I rank seventeenth really will be hurt. On the other, the Condorcet paradox disappears. If we use the Borda procedure to rank A, B, and C, there can be no inconsistency: if the vote shows A ahead of B and B ahead of C, then it will show A ahead of C. So it may seems that we have found what we were after, namely a consistent way for assemblies to express their will. Unfortunately, that is not so, for there are other drawbacks to the Borda procedure. Imagine, for instance, that two candidates, A and B, stand for election. There are 30 voters in the electoral college, 19 of whom prefer B to A, so it would seem that B will carry the day. If the Borda procedure is used, however, the supporters of A can rig the election by introducing a third candidate, C, who is universally disliked. So the supporters of B will rank C third and last, but the supporters of A will rank C second, against their own preferences, for the sole purpose of defeating B. Then A will be ranked first 11 times and second 19 times, for a Borda score of $11 + 38 = 49$, whereas B will be ranked first 19 times and third 11 times, for a Borda score of $19 + 33 = 52$, so A gets elected.

There has been much research since on voting procedures, culminating in the so-called Arrow impossibility theorem, which states that there is no procedure that is immune both to Condorcet's paradox and to this kind of manipulation. The only way an assembly can get around them is to appoint a dictator and defer to his decisions. In other words, there is no perfect procedure: the problem is to choose the best one for the circumstances.

In collective decision-making, the outcome is decided by the voting procedure as much as by the voters' preferences. This is a lesson that politicians have known for a long time; every representative is aware that using the rules book and scheduling the votes in the right order can perform miracles. As a historical example, let us recall that on June 20, 1991, the German Bundestag had to pick one of three options: A, transferring the government and the Parliament to Berlin; B, leaving both in Bonn; and C, leaving the government in Bonn but transferring the Parliament to Berlin. It is now clear that there was a substantial relative majority for staying in Bonn, so that putting all three choices to the vote together, and either using the relative majority or the Borda procedure, would have led to that outcome, for most of those who favored moving everything to Berlin preferred staying in Bonn to separating government from Parliament.[1] However, the procedure was deferred to a committee, who decided that the assembly would first vote on the compromise position C, and if it did not get a majority, they would then choose between A and B. This freed the voters from that concern, and the outcome, as we well know, was that everything moved to Berlin, a move of historical importance, which ultimately hinged on the voting procedure rather than on some sense of the public interest or the common good.

There is, however, one last branch to cling to: certainly if *every* member of society prefers A to B, then society itself should prefer A to B. This is called the Pareto criterion, from the name of Vilfredo Pareto, an Italian sociologist and economist (1848–1923), and it is met by all the collective decisions rules we have discussed: the majority vote, the Borda rule, and even the dictatorial rule (if every member of society prefers A to B, then so does the appointed dictator, and he will accordingly choose A over B). Unfortunately, this criterion may not be strong enough to choose between two alternatives: what are we to do if some of us prefer A and others prefer B? We will have to resort to some form of collective decision-making, and we are back with the difficulties we outlined before. On the other hand, the Pareto criterion will enable us to eliminate many inferior outcomes: if everyone prefers A to B, and C to D, then there is no need to even consider B and D, and the real choice is between A and C.

The Pareto criterion is a criterion for efficiency: if A and B are two possible states of society, and if A is Pareto superior to B (that is, if everyone prefers A to B), then A is a more efficient use of the collective resources. Put another way, if society is in a certain state, and there is another possible state which everyone would prefer, then society must be wasting some resources. For instance, imagine sharing a cake. There are many possible ways to do so, from the dictatorial one (give everything to a single person), to the most egalitarian (give an equal share to everybody). We cannot rank them by the Pareto criterion: all of them turn out to be

1. W. Leininger, "The Fatal Vote: Berlin versus Bonn," *FinanzArchiv* N.f. 50.1 (1993): 1–19.

efficient. For instance, sharing equally will not be unanimously preferred to giving everything to a single person, for the lone beneficiary will dissent. What the Pareto criterion tells us is that it is inefficient to leave some of the cake undistributed, for if this happened, we could distribute the leftovers and give a little bit more to everyone.

Achieving economic efficiency is not always that simple: it is difficult to track down waste and slack in large organizations. For this reason, many economists advocate worrying about efficiency first and letting redistribution take care of itself. In economic development, for instance, this means applying policies that will increase the GNP, in the hope that this global increase will end up benefiting everyone, perhaps through an unspecified "trickle-down" process from the rich to the poor. Unfortunately, there is no convincing reason why it should be so, and there is historical evidence for the opposite. The Russian economy, for instance, certainly is more efficient than it was under the Soviet regime, but large segments of the population, such as the pensioners, are worse off than they were before the fall of communism. In addition, the argument of efficiency has often been used for political purposes. The whole process of colonization, for instance, has consisted of Europeans settling into foreign lands and dispossessing or exterminating the natives under the pretext that they would put that land to better use than its former owners did. Underdeveloped countries, mainly former European colonies who have learned their lesson the hard way, are understandably wary of entering global trade agreements, even though they would increase the efficiency of the world economy: they worry about redistribution.

All this means that the Pareto criterion is not a suitable one for optimization, because it is not discriminating, and may be used to justify extremely unfair situations. If we are to use optimization theory as a tool for economic decision-making, we need to find another criterion. In such a simple matter as deciding whether to build a new road, we run into a host of problems. If that road is built, there will be winners (those whose travel routes have been shortened, businesses that will spring up along the new road), as well as losers (property owners whose land is taken up by the new road, businesses who lose customers as traffic patterns change), both probably very vocal. How are we to balance the wishes of one against the other, bearing in mind that some of the parties will not be around to express their feelings? Granted that building the road is deemed globally beneficial, would there not be a better use for that money, for instance by building somewhere else a road that would be even more beneficial? What about scrapping the road and putting the money into education or health care? The Pareto criterion does not tell us which of these projects to choose. All it says is, whatever you do, make sure you do not waste.

So we need some criterion to help us decide. There are several possible ones, and the one you choose will represent your view of the public interest, or the common good. One of the most popular criteria consists of having each of the parties affected by an economic project evaluate the benefit or the prejudice that realizing that project would bring him or her; one then simply sums up the benefits and subtracts the costs from realizing the project and the prejudices borne by individuals. This is called the *utilitarian* criterion, and the result is called the *social value* of the project. If it is positive, the project is deemed to be in the public interest, which does not mean it should be realized, because there could be another project, somewhere else, with an even higher social value. In fact, the main problem is not so much finding which projects have positive social value, as finding the optimal ones, those with the highest possible social value.

The utilitarian criterion is basically a monetary version of the Borda rule: each stakeholder votes for or against the project by telling how much he or she would benefit or lose from it. Although this is one of the most widely used criteria used in public economics, it shows some weaknesses. First of all, it is not very satisfactory to express everything in monetary terms. Both a wheat field and a nuclear plant carry capital value, but they are also means of subsistence for individuals, and if they are deprived of it, one will have to evaluate the costs of relocating and changing jobs. In addition, evaluating environmental costs is going to be very tricky as well: how are we going to find out the price tag individuals put on clean air and quiet nights? Not only is it a difficult matter per se, it is also open to all kinds of false pretenses and manipulations: one is always tempted to exaggerate one's prejudice, so as to be able to claim more compensation, and to keep one's true benefits hidden from the tax collector. And even if I am honest and willing, why should I be the best judge in my own cause? What if I have expensive tastes, which I have acquired by squandering vast amounts of unearned wealth? What if I have willingly taken risks from which it is now very costly to extract me? Certainly, if I am very rich, the amount of money I would regard as appropriate to compensate me for suffering some inconvenience would be much greater than if I am very poor; a sleepless night is worth much more to Bill Gates than to Mother Theresa. What should we do about these differences?

In John Rawls's famous book *A Theory of Justice*, we find another criterion for the common good. To compare two possible states of society, A and B, let us look at how the worst off fare in each of these two states.

Let us say that, in state A, group 1 is the worst off of all members of society: their level of income, say, is i_1. In state B, group 2 is the worst off: their level of income is i_2. We shall say that state A is better than state B if i_1 is higher than i_2. In other words, a state of society is preferable to another if the poorest members of that society (which may differ from one state to another) are treated better in the first state than in the second. To be fair to Rawls, his criterion comes second to other considerations: he first asks society to respect certain basic rights. If these are fulfilled, economic considerations come into play, and then he classifies possible states by using this criterion. Note that between the utilitarian criterion and the Rawlsian one, there are many others, obtained by giving different weight to individuals when computing the social value of a project. For instance, if we decide that every poor citizen will count as much as two rich ones, we will multiply his benefit (or his prejudice) by two before carrying it into the total balance. In this way, more consideration is given to the effects of a given project on the poor, and in the Rawlsian criterion, they are the only ones to be considered.

There is no natural choice between all these criteria: the one you pick defines your idea of what the public interest is. If you are a utilitarian, you pick the first one; if you are a Rawlsian, you pick the second. You can also define your own: if you feel that voters in your constituency, or contributors to your reelection campaign, deserve more attention, you simply give them more weight in computing the social value of projects. This is a crucial step in social planning, for the criterion one chooses embodies a compromise between all the different objectives to be achieved. Once this hurdle is cleared, however, we will find it is only the first one: we quickly run into another obstacle, which is the question of implementation. Having formulated the decision to be made as an optimization problem and having found the right solution are no longer enough; the social planner has to show how this solution is to be implemented. In a way, the engineer encounters that problem as well: it is not enough to find the right shape and composition for the bridge; one also has to explain how to build it out of existing machinery, or to construct a new machine. But the engineer deals with machines and materials, whereas the social planner deals with human beings.

Much of the progress in economic theory in the past thirty years has been devoted to studying the problems that social planners face. They fall roughly into two categories: asymmetry of information, and strategic behavior of individuals. If we assume these problems away, we get a wholly unrealistic picture of how an administration is run. Unfortunately, it is still prevalent in political thought, at least in France. Every civil servant is supposed to do her job perfectly well, out of sheer love for the public, in the absence of monetary rewards for performance. Every mayor, police officer, or judge, although endowed with largely discretionary powers, is supposed to stop short of using them for personal purposes, out of respect for his duties. Every business is supposed to promptly provide the administration with all information it needs to compute taxes, particularly environmental taxes for activities generating negative externalities. The judicial system is supposed to be so effective as to prevent any kind of collusion, corruption, or favoritism when the administration awards contracts for public works. It is also assumed that government policies directed toward supporting certain categories of citizens will never be diverted toward other categories, who were not originally among the intended beneficiaries but who would claim to be.

I am not claiming that civil servants or public officials are corrupt or do not care about the public interest, I am just saying that, like everyone else, they have their own view of what that interest is. Generals will tend to believe that more money should be spent on defense, and an automobile manufacturer is famously quoted as saying "What is good for General Motors is good for the United States." Everyone tends to speak from experience and see the public interest from his own position in society; it is difficult to factor in facts that you have no direct experience of, and ideas which you are not familiar with. In addition, power certainly may fall into the hands of individuals who will try to use it to their own particular ends; indeed, one may fear that precisely this kind of person will be attracted to a position of responsibility, and will try to increase her share of power. The traditional remedy is to create some watchdogs, individuals or institutions who will make sure that the name of the public interest is not taken in vain, and that individual ambitions are kept under control. One then runs into the age-old problem: who will watch the watchdogs? Are we to suppose that they will be immune to human weaknesses, and will remain forever uncorrupted by the exercise of power?

If we assume that, somewhere in the system, there is a person or a group that is entirely devoted to our idea of the public interest, we may just put them in charge and let them govern the ship of state. This was the basic idea of many utopian states, such as Plato's ideal republic, which was ruled by philosophers, or the Soviet Union, which was run by the Communist Party. Plato was lucky enough not to have his ideas put to the test, but the Soviet Union turned into a murderous tyranny, instead of the socialist paradise its founders were hoping for. This is but a particular aspect of a more general problem: institutions may be designed with a specific purpose in view, and perform quite differently than it was intended. A famous instance is the Fourth Crusade, launched throughout Christendom to free the tomb of Christ from the heathen. Instead of going to Jerusalem, it ended up under the walls of Constantinople, the capital of the Byzantine Empire, which it took in 1204 in an orgy of massacre and pillage against fellow Christians. There is no doubt that this expedition, far from strengthening Christianity in the East, hastened the decline of the Byzantine Empire and its eventual conquest by the Ottoman Turks.

Without even appealing to historical examples, any one of us can recall instances to show that organizations are not animated with a single purpose, but that their members constantly jockey for power and try to steer them in directions that will further their own careers and ambitions. Wars have been fought, and men sent to slaughter, for very mundane interests. The lesson of experience is that it is useless to conceive institutions which will be directed by providence and staffed by angels, able to read hearts and minds and entirely devoted to their fellow citizen's welfare. In the real world, institutions will function with very ordinary men and women, aware of their duties, but also preoccupied with their own careers. They may be doing their best under the circumstances, but their knowledge will always be fundamentally limited. Relevant information will be hidden from them, they will have only that much time to devote to each piece of business, and they will have to develop routines which will harden them and channel their thought process. Every day will bring them in contact with other individuals and organizations with their own objectives and preoccupations, and these regular exchanges will lead to a mutual adjustment, like the pebbles on a beach are uniformly rounded by the waves and tides scraping them together.

Economists are often faulted with being too optimistic in certain ways, and too pessimistic in others. Their models are peopled with individuals who are farsighted and selfish, with an endless ability to compute. These creatures are able to foresee the consequences of their actions, and the consequences of the consequences. They will never tire of figuring things out; their intelligence easily runs up and down the causality ladder; and neither passion nor impatience ever interferes with the limpidity of their reasoning. They know that everyone else in the world they inhabit is made on the same pattern, and they act on that information, as on any other they have. They take into account the projected response to their own actions: knowing, as they do, that others are similar to them, they can figure out their response by putting themselves in the others' shoes, much like a chess player trying to look a few moves ahead. The way organizations work, be it private firms or government, is similar to a huge chess or poker game, where each player is trying to adapt her own behavior to what she can infer from the perceived behavior of others. Since everyone is doing the same, we may have a situation where everyone does exactly what the others expect of him, and ends up having acted in his own best interests, given that the others have done exactly what he expected them to do. Such a situation is called an *equilibrium*, and was described at length in chapter 7.

One may suspect that the whole equilibrium concept is unrealistic: do we really act rationally? Do we think out the consequences of our actions? Are we not guided mostly by momentary impulses, the force of circumstances, or the power of habit? It turns out that this is just one particular way to look at equilibrium. We have introduced it as a situation resulting from rational beings arriving simultaneously at the same conclusions. But it can also be the end result of a simultaneous learning process from agents who have just enough sense to realize whether their previous policies have been successful, to keep applying them if they have, and to change them if they have not. We have also described earlier how species may adjust to each other and to the environment, so that individuals may be born into an equilibrium and support it by innate behavior, rather than intelligent strategies.

An equilibrium is to a large extent artificial, for it depends as much on the expectations of the parties concerned as on any objective component of the overall situation, much as the price of a stock or the rate of a currency depends as much on what the administrators think as on any fundamentals of the firm or the country. There can be many different equilibria, each of them being a particular way of coordinating expectations and actions, so that the first support the second. Each of them is a situation where the expectations that I entertain about everyone else determine a certain strategy on my part, and on everyone else's as well, and where all these strategies result in precisely everyone's expectations being confirmed. In other words, if a group is in equilibrium, its members are never surprised by each other's behavior. To outsiders, it appears as a set of conventions that rule social interaction and which everyone adheres to; these conventions are self-fulfilling, in the sense that, given that others conform to them, our best interest is to conform as well.

The point, however, is that there is nothing natural about such conventions. For starters, mathematically speaking, many different equilibria are possible. Montaigne, and many others after him, made a point of comparing the customs and ethics of different peoples and showing how different they are from the ones our own education takes for granted.

The purpose of education is to train us to live in a particular society, that is, to develop in us a set of habits and values according to which the equilibrium we are born in is the only "natural" one, and "morally superior" to the others. We feel it to be natural because it is the only one in which we are comfortable, and we think it to be morally superior because we have not been trained to share the values the other equilibria convey. This is not to say that there is any worthier alternative. Social conventions are unspoken agreements that make society livable by indicating to everyone what role to play, and what to expect from others. We are mostly born into our roles, and there is little point in wondering how that particular casting was arrived at. The important point is that it has been done, and that everyone understands his or her role.

As Pascal puts it, "How right one is to tell men apart by their exterior appearance, rather than by their interior qualities! Who will have precedence between the two of us? Who will give way? The least talented? But I have as much talent as he has, and we will have to fight over it. He has four footmen, and I have only one: this is clear; we just have to count; it is for me to yield, and I would be a fool to discuss that matter. By this simple device we have achieved peace, which is the greatest of all goods."[2] Similarly, Pascal points out that law is but a matter of convention. The law draws its authority from the simple fact that it is the law, and is recognized as such; there is no point in claiming that it draws its legitimacy from some divine authority or some natural law. The law has to be obeyed because it is the law, not because it is just: "Custom makes fairness, for the lone reason that it is accepted; this is the mystical foundation of its authority. Whoever brings it back to its principle destroys it. Nothing is as wrong as laws which redress wrongs; for whoever obeys them because he thinks they are just obeys the justice he imagines rather than the essence of the law: it is entirely contained in itself. It is the law, and nothing more. Whoever will look into its foundation will find it so weak and slight that, unless he has grown used to the prodigies of human imagination, he will marvel that one century has been enough to coat it with such pomp and reverence."[3]

We have now come full circle. We started this chapter by wondering whether the tools of optimization theory could be used for regulating human society, and we discovered that we needed a criterion to optimize, that is, an acceptable definition of the common good (or the public interest). Lacking such a definition, we started on another tack. Each

2. *Thoughts*, frag. 320.
3. Ibid., frag. 230.

individual in a group is supposed to know where his or her best interests lie. Supposing they are rational, that is, that they act consistently and strategically to further those interests, where does that lead the group? The answer is that it leads it into some equilibrium, but there may be many of these, and how are we to choose between them? For this, we need a criterion for the common good, and this is precisely what we have been missing from the beginning.

One can, of course, impose such a criterion, or educate a group into accepting it. For instance, some equilibria might turn out to be inefficient, so that shifting society into another equilibrium, and redistributing the wealth thus created, may be to everyone's benefit. Even such a seemingly straightforward operation may be quite difficult to perform: people will lie about their own situations, thereby skewing the process, and it will be no easy matter to ensure that the redistribution actually takes place. There are other cases when the common interest is less clear. In the case of global warming, for instance, there is by now a consensus in the scientific community that maintaining human production of greenhouse gases at their current level will lead to climate change early in the current century, with catastrophic consequences for most countries. Bangladesh would simply disappear, coastlines would recede everywhere, Europe would lose its mild winters and become similar to the eastern United States. Other countries, however, may fare better; with the disappearance of the permafrost, Russia may gain vast amounts of arable land in Siberia. Most of us would think that the general interest lies in preventing this from occurring, and therefore curtailing right now the production of greenhouse gases. This, however, is perceived as extremely unfair by developing countries, which are not responsible for this situation and are now asked to pay for it in terms of postponed economic growth. It is also perceived as unfair by the principal beneficiaries of the present situation, the United States, who have come to see their way of life as the only "natural" one, and who see no reason why it should be curtailed, even though it is unsustainable for the rest of the planet. There are essentially two possible outcomes for such a situation. Either there will be some kind of general agreement on reducing the emission of greenhouse gases, with an international authority given the power to monitor compliance, or there will be a revival of colonization from the United States and Europe, who will seize desirable tracts of land in the new geography and maintain emerging powers like China and India in economic underdevelopment. The first solution means implementing a certain idea of the common good; the second one is just an equilibrium, backed by military force, like so many others before it. The political developments of the past years can be read as a struggle between the proponents of these two options.

We have now reached the end of our journey. It started in the world of the Renaissance, impregnated with Christian values. It is impossible to understand thinkers like Galileo or Leibniz, for instance, if one does not understand that they believed the world to have been created by God. The laws of nature then are simply the rules God followed when creating the world, and the purpose of science is to recover them from observations. There is then also a deeper science, which is to seek the purpose God himself had in creating the world. This is what Maupertuis, in a glorious moment, thought he had achieved, thereby reconciling forever science and religion, both being the quest for God's will, in the physical world and in the moral one.

Our journey ends in a world where God has receded, leaving humankind alone in a world not of its choosing. Technological progress, however, has enabled us to play God, by shaping our environment and ourselves, on a scale which has now reached the planet and is growing at an unprecedented pace. What do we want to do with this power? What kind of world do we want to create, among the many possible ones? This is an entirely new question, which humanity faces in an entirely new situation. Our intellectual categories and moral values, which were developed in earlier times, have yet to incorporate the changes that science has brought in the human condition. Our personal identities and characters, for instance, can be changed by chemical treatment; our bodies can be changed by cosmetic surgery. Our tastes and opinions are shaped by professionals and spin doctors, from marketing, advertising, mass communication, and journalism. Huge amounts of money and ingenuity are invested into making us desire what industry wants to sell and approve what people in power want to do. What then is the meaning of the traditional advice of moral philosophy, "know thyself"? Should I meditate the consequences of being depressed and nearsighted, or should I take Prozac and seek eye surgery? It is claimed that Beethoven died of lead poisoning, and that it also stimulated his creative powers. Should he have been cured? Who is the real Beethoven? The deaf genius who wrote the Diabelli Variations and the Ninth Symphony, or the more ordinary musician but healthy one who would have replaced him if his condition had been diagnosed, as it certainly would have been today?

With the progress of genetic engineering, the human species may soon prove to be as malleable as the individual. We can already detect undesirable genes in the fetus, thereby raising moral problems where there were none before. Between screening the genes and actually changing them, that is, shaping our offspring according to our wishes, there are but a few decades of technological progress. The time is coming when plain people, ordinary couples wanting a baby, will face a task that philosophers, theologians, and moralists of ages past believed to be beyond our powers: designing a human being. There is as yet little precedent for this kind of decision, apart from the book of Genesis and other myths of Creation, but as it becomes part of the human experience, guidelines will develop, one way or another. Whatever they are, they will leave open to the human species the possibility of reshaping itself, of taking command of the evolutionary process. This is the last blow dealt to the idea of the common good, as we understand it today. It is hard enough to define the common good of a society of rational individuals with well-defined objectives and tastes. But if the individuals alive today are but stepping stones toward a higher and better state of humanity, of which we know nothing yet beyond the fact that they are different from us, then the task becomes impossible, like building a statue out of water.

(CHAPTER 10) A Personal Conclusion

THE STORY I HAVE TOLD in this book is not one of failure. On the contrary, it is a record of tremendous and unexpected successes. Only four hundred years separate Hubble, cruising several thousand miles above the Earth, and sending us pictures from galaxies billions of light-years away, from the primitive telescope which Galileo used to discover mountains and seas on the Moon and rings around Saturn. The modern-day physicist no longer climbs towers to drop stones, but sends sub-atomic particles colliding against each other in circular accelerators many miles in diameter. Galileo's original law on the motion of falling bodies (speed increases in proportion to time elapsed) is now seen as a minor consequence of much deeper properties of space and time, encapsulated in Einstein's theory of general relativity.

Maupertuis thought that the least action principle was the blueprint of Creation. On the one hand, it contained all the secrets of nature, since the laws of physics could be derived from it by mathematical arguments. On the other, it had such an obvious purpose that it was clear that there was a will behind it. How can natural motions minimize the total ex-penditure of "action" if there is not an invisible hand to guide them? How does light choose the shortest path among the many possible ones if some higher intelligence does not direct it, like a skier tracing a path in virgin snow for others to follow? We now know that this is not the case, and that reality is much richer than that. Light does not follow the shortest path; natural motions do not minimize the action. The right concept here is stationarity rather than minimization, and it requires some mathematical sophistication to understand. So there is seemingly a loss, since the concept is one more step removed from our everyday experience, although it is more than compensated by its increased ap-plicability. Today's physics and mathematics are much richer and broader than those of Maupertuis, but the least action principle (in its revised version, with action-minimizing motions replaced by stationary ones) remains in force. It is no longer considered a fundamental law of nature, but a mathematical tool toward new discoveries—such as Gromov's uncertainty principle. There is no need either for an invisible hand to help light find the stationary paths: it is but one of the many consequences of the fact that light consists of waves, which propagate and interfere according to laws which were already known to Huygens, many years before Maupertuis. In a similar fashion, the discovery of quantum physics provided a firm foundation for the least action principle: it is but a macroscopic consequence of the structure of matter at very small scales, as Feynman pointed out sixty years ago.

The least action principle ended up looking very different, but more interesting, than Maupertuis had suspected. The metaphysics have gone, but the physics and the mathematics are deeper and better. Perhaps Maupertuis would have been disappointed by the way it turned out, but the greater scientists, Fermat, Huygens, Euler, Lagrange, would have been thrilled at the progress we have made in understanding nature. Unfortunately, they would probably have expected the development of human society to match the progress of scientific knowledge, and they would have to be sorely disappointed. If they were alive today, they would have to be told about the horrors of the twentieth century, the tens of millions killed in battle during the two world wars, two cities wiped out by a single atomic bomb with no warning to civilians, the millions of tons of bombs dropped over Germany and Vietnam, the millions of displaced people in Europe, the Middle East, and Africa, the genocides of the Armenians, the Jews of Europe, Cambodia, and Rwanda. Killing people has now become an industrial process like any other, benefiting from advanced technology and management skills, and the efficiency of that particular industry has kept pace with the rest. It looks like the species *Homo sapiens* has not evolved since it first gathered into competing tribes, but that they are now throwing bombs at each other instead of rushing at the enemy with spears. Worse still, those who manage the killing process are so remote from the end result that they do not feel that they are doing something wrong or even unusual. They have desk jobs, like Adolf Eichmann had; in the words of Hannah Arendt, evil has become banal.

This discrepancy between progress in technology and the persisting mistreatment of human beings is very troubling, even for those who have not been on the receiving end of humanity's malevolent ingenuity. One naively expects that the enormous stock of knowledge that has been acquired in the past four hundred years, and that has enabled some of our species to tread on the moon, would also enable us to live on this planet in abundance and peace. This is obviously not the case. Europe, that most cultured of continents, where pacifists and socialists have always abounded, started two world wars and exterminated most of its Jews. We have not learned from this. All my adult life, torture has been used as a way to terrorize people, by dictatorships in Latin America, Africa, and the Middle East, but also by great democracies, France, Israel and the United States, and I find this deeply unsettling: not only have we more efficient means of killing each other, we also have more efficient means of inflicting pain and taking every advantage of it, so that killing may look merciful by comparison. Since 9/11 the situation has gotten worse; even in Great Britain, the birthplace of habeas corpus, one can now be detained without knowing what one is accused of or who one's accusers are. In the international sphere, the U.S. government is walking away from the complex network of international agreements, starting with the United Nations charter, which tried to create some kind of international rule of law, and proclaiming instead its right to preemptive strike, that is, immediate military action against anyone by whom it feels threatened, anywhere in the world. This is the kind of doctrine that was entertained by the Roman Empire; the natural sciences have progressed tremendously in two thousand years, while political science clearly has not.

One way to react to this sorry situation is to despair: what good is scientific knowledge if it puts stronger weapons in the hands of the powerful? What achievement of quantum physics can atone for its inventing the atomic bombs? There are the lasers, the CD and DVD players, all the digital technology that is now so central to our lives, and all the peaceful uses of atomic energy which may save us when the fossil fuels have run out or we have finally decided they are too dangerous to burn. True enough, but even if today people live longer than their ancestors, are in better health, enjoy more comfort of every kind, they may not be happier. This is because happiness is relative to one's experience and to the people one consorts with; if I feel depressed, I will not be comforted by the thought that two centuries ago, few people reached my age. It is also because the mass media of communication have turned the world into a global village, and are bringing into every household vivid pictures of incredible hardships. Famines and massacres have always happened throughout the world and the centuries, but only now are they brought to our immediate attention by the information networks and the Internet. The human experience is now global, as is the economy. It is largely shaped by newspapers, radio, and television, and our sense of comfort and happiness depends on what we read, hear, and see. The end result is that we are directly exposed to many more tragic situations than our ancestors, that we see these situations, like the war in Iraq, perpetuating without improvement, and that we therefore tend to view the world as much less of a safe and happy place.

There is a general feeling that science has given us longer and better lives, but has not taught us how to live them. Certainly, we have witnessed immense progress in very many fields of knowledge, from mathematics to anthropology, and one could even argue that in effect, science has split into separate sciences, each one with its own set of operating rules, and each one with its record of successes. However, no unified view of the world has emerged. In fact, most scientists are extremely specialized, and have very little to say about science outside their own fields, let alone about the world at large. One can find scientists of practically every opinion and creed, and scientists have been directly connected with very questionable enterprises. There is a sense of disappointment that people who have been able to devise the means to reach the Moon are not able to answer the basic questions that every human being faces: who am I? what am I to do? Science seems to raise more questions than it provides answers, but human beings are in quest of certainties, and if science will not provide them, then others will—religions and ideologies. And indeed, the first half of the last century was the era of ideologies, which ended with the bloody clash of fascism and communism, while the second half has seen religions emerge as the main actors, and may yet lead us to another conflagration between the Abrahamic creeds—the so-called clash of civilizations.

I think these are the wrong attitudes. There is no cause for despair, nor should we let religious fundamentalisms lead us down the path to collective destruction. Since the time Galileo first raised his telescope to the night sky, we have learned much more than how to send humans to the Moon: we have learned a method of investigation. It consists of relying on facts, and of arguing correctly from them. Establishing the facts is an aim in itself: ideally, one should put in doubt all the certainties handed down by tradition and society, and try to find, by repeated observation and experimentation, some truths one can safely build upon, truths which are to be perpetually revisited and double-checked. This is the scientific method, as theorized by René Descartes; as indicated in the title of his 1637 book *A Discourse on the Method to Orient One's Reason and to Seek Truth in Sciences*, this method is universal. It has been used in science with tremendous success, and there is no reason why it should not be as useful in philosophy, or in trying to establish some principles by which to guide our collective and individual lives. This is precisely what Descartes tried to do, and this is what we should be doing, each in our own way, reexamining what we believe in, in the light of our experience. Some of the conclusions Descartes reached we would reach ourselves. He realized, and accepted, that his method did not lead to any specific morals or ethics, much as we see nowadays that science does not teach us how to live our lives. But he introduced the idea of a *morale par provision*, "acting morals," so to say, to be in effect until better ones are found, much like a scientific theory is in effect until progress has made it obsolete in favor of another one. In other words, in morals as in science, we have not reached a definitive and all-embracing truth, nor is it certain that we will ever reach one. But the history of the past centuries has shown us that incomplete and provisional truths can be used to great effect in science, and there is no reason it should be any different in morals and philosophy. In other words, do not be disappointed that we do not have right now the definitive answer to problems humans have been struggling with since the beginning of time. It is enough to have partial answers, to cobble them together in an incomplete but operating theory, and to work toward making it more complete.

What I am claiming here is that rationalism, the reluctance to accept anything that cannot be defended by argument or experience, has not exhausted its possibilities and can still carry us a long way. In a way, this is just another belief, and it is irrational as all beliefs are. I myself think that this faith in reason is supported by experience, whereas faith in resurrection or reincarnation is not, but this argument will not convert a Christian or a Hindu unless she is ready to submit to experience, that is, unless she already is a rationalist. Why then put our confidence in rationalism and the scientific method?

A first set of reasons has to do with the alternatives. If we do not accept that reason has to be the mainspring of human action, that our decisions have to reached by rational arguments with due regard to their consequences, then we must let emotions and passions take their toll. It is true that there is an irrational side to human beings, which can bring out the best in them, but it can also bring out the worst, and in the long run it is the worst which prevails. If rational arguments are not accepted, disputes will be solved by force, if no common ground is accepted for discussion, then the only recourse is violence. As Popper puts it, "Rationalism is closely connected with the belief in the unity of mankind. Irrationalism, which is not bound by any rules of consistency, may be combined with any kind of belief, including a belief in the brotherhood of man; but the fact that it can easily be combined with a very different belief, and especially the fact that it lends itself easily to the support of a romantic belief in the existence of an elect body, in the division of men into leaders and led, into natural masters and natural slaves, shows clearly that a moral decision is involved in the choice between it and a critical rationalism."[1] Popper goes on to say that

> faith in reason is not only faith in our reason, but also—and even more—in that of others. Thus a rationalist, even if he believes himself to be intellectually superior to others, will reject all claims to authority since he is aware that, if his intelligence is superior to that of others (which is hard for him to judge), it is so only in so far as he is capable of learning from criticism as well as from his own and other peoples' mistakes, and that one can learn in this sense only if one takes others and their arguments seriously. Rationalism is therefore bound up with the idea that the other fellow has a right to be heard, and to defend his arguments. . . . Ultimately, in this way, rationalism is linked up with the recognition of the necessity of social institutions to protect freedom of criticism, freedom of thought, and thus the freedom of men."

My own understanding is that the ability to argue rationally, and hence the ability to create and understand science as we know it, is precisely what distinguishes human beings from other animals. An interesting situation may arise if, in the future, we encounter creatures who behave in some pattern which is incomprehensible to us, and yet appear to use nature in a way which suits their (unfathomable) purposes; such a situation is described by Stanislaw Lem in his book *Solaris* (where the creature actually is a whole planet), which deals with human intruders much as we would deal with a colony of ants, trying to figure out how they react to stimuli and how they communicate. But this is science fiction, and although it does bring up very deep problems about the very concept of rationality, humanity has more pressing concerns right now. Again, I think that human beings can and should recognize each other in their capability to think rationally—even if they think and behave irrationally in many circumstances—and that is the strongest bond among them. This capacity is apparent even in societies who were never exposed to science as we know it. As Claude Lévi-Strauss has repeatedly pointed out, human beings string together intellectual systems from whatever bits and pieces are at their disposal, much like an amateur handyman, who does not have the same tools as a professional (and who may not even know that they exist), will make do with whatever tools he

1. Karl Popper, *The Open Society and Its Enemies*, 4th ed. rev. (Princeton, NJ: Princeton University Press, 1962), chap. 24.

has at hand and turn them to his needs. The variety of human experience must be put into some semblance of order, and if science is not available for that purpose, then societies turn to myths, religions, and ideologies. Grooming such systems of beliefs, keeping them alive by continuous additions, devising ways to connect their various parts and to smooth over discrepancies, not to mention contradictions, requires considerable intellectual ingenuity. If one thinks, for instance, of the intellectual power required to write the *Book of Safety* by Ibn Sina (Avicenna), the *Guide of the Perplexed* by Moses Maimonides, or the *Summa Theologica* by Thomas Aquinas, all of which try to build a system of beliefs reconciling the Abrahamic religions with Aristotelian philosophy, one cannot but put their authors at the highest level of intellectual achievement. Avicenna and Maimonides were first and foremost known as physicians, and most of their work deals with medicine; they were in fact the greatest scientists of their time. If they were alive today, they would know of the evolution of species and the genetic code; one wonders what kind of system they would develop from this vastly increased store of knowledge.

We should proceed in the direction they and so many great minds have pointed out to the following generations: use the power of reason to free people from all kinds of bondage, bondage to natural powers and bondage to human oppression. What is needed is courage: it is always so much easier to accept what you are being told than to think for yourself. "Laziness and cowardice," writes Immanuel Kant, "are the reasons why so many men, after nature has so long freed them from a foreign conduct, nevertheless remain all their lives in a state of tutelage; and why it is so easy for others to pose as tutors."[2] But the remedy is at hand; as Georg Christoph Lichtenberg puts it, "That is true, gentlemen, I cannot make my own shoes, but my philosophy, I will let no one pick it for me." Deciding on one's own opinions, even in important matters, is not such a difficult task as all those who like to pick them for you would want you to believe. In recent times, Noam Chomsky has made that point forcefully, as, for instance, here: "I will not connect the analysis of social issues with scientific problems, which require specialized and technical training and intellectual references before they can be treated. To analyse ideology, it is enough to have a good look at the facts and to be willing to follow an argument. Only Cartesian sense, 'the most common thing in the world,' is required. . . . It is Descartes' scientific approach—if by this you mean to be willing to look at facts with an open mind, to check the assumptions

2. Kant, "Was heisst: sich im Denken orientiren?" *Berlinische Monatsschrift* 8 (July-December, 1786).

and to follow an argument to its conclusion. Nothing further is required, no esoteric knowledge to explore 'depths' which are simply non-existent."[3]

Of course, those who subvert the power of government for their private interests tend to protect themselves by shielding public policy from scrutiny. The traditional way to do this is to endow rulers with some suprahuman legitimacy, to picture them as God's vicars on Earth, as the defenders of morality or as the leaders of a nation at war, so that any criticism of their actions can be represented as a criticism of the higher values they stand for. A more subtle way to achieve the same aim is to pretend that global issues are much above the head of common people, either because they require some deep and technical knowledge which is available only to experts, or because ordinary people do not think as clearly as politicians, and do not have as deep a commitment to the common good. All of this is patently false: there is nothing sophisticated about understanding global warming, and ordinary citizens are much more concerned about its consequences than politicians. As a third method, it is also very effective to hide private interests under noble ideals. Now as always, armies have been sent to invade land and to seize resources from weaker peoples, always for the sake of religion or civilization, never it seems out of greed. The motives provided—in earlier times, we wanted to save the souls of the heathen; nowadays we want to bring them democracy and deliver them from oppressive regimes—are a tribute to the fertile imagination of the human mind, and to our unending capacity to hide the reality of our actions under a stream of words. Sometimes I feel like Robert Musil's hero in *The Man without Qualities*, who saw "the big sentiments, ideals, religions, fate, humanity, virtue, as the ultimate evil. He ascribed to them the fact that our times were so insensitive, so materialist, so irreligious, so inhumane and so depraved."

Big sentiments are no guarantee of ethical behavior, as the behavior of troops brought in to convert the heathen to our religion or our way of life abundantly shows. In the realm of moral laws, as in the realm of natural laws, the scientific method is the only safe one. Here again, let me quote Popper at length:

> On the contrary, whenever we are faced with a moral decision of a more abstract kind, it is most helpful to analyse carefully the consequences which are likely to result from the alternatives between which we have to choose. For only if we can visualize these consequences in a concrete and practical way do we really know what our decision is about; otherwise, we decide

3. *Dialogues avec Mitsou Ronat* (Paris: Flammarion, 1977).

blindly. In order to illustrate this point, I may quote a passage from Shaw's *Saint Joan*. The speaker is the Chaplain; he has stubbornly demanded Joan's death; but when he sees her at the stake, he breaks down. "I meant no harm, I did not know what it would be like ... I did not know what I was doing ... If I had known, I would have torn her from their hands. You don't know: you haven't seen: it is so easy to talk when you don't know. You madden yourself with words. But when it is brought home to you; when you see the thing that you have done; when it is blinding your eyes, stifling your nostrils, tearing your heart, then, then—O God, take away this sight from me!" There were, of course, other figures in Shaw's play who knew exactly what they were doing, and yet decided to do it, and who did not regret it afterward. Some people dislike seeing their fellow men burning at the stake, and others do not. This point (which was neglected by many Victorian optimists) is important, for it shows that a rational analysis of the consequences of the decision does not make the decision rational; it is always we who decide. But an analysis of the concrete consequences, and their clear realisation in what we call our "imagination," makes the difference between a blind decision and a decision made with open eyes; and since we use our imagination very little, we too often decide blindly. This is especially true if we are intoxicated with an oracular philosophy, one of the most powerful means of maddening ourselves with words—to use Shaw's expression.[4]

Seek the truth, and the truth will set you free. This is a pronouncement as old as philosophy itself, but science has taught us what truth is really like. It is not a global all-encompassing truth, handed down by some supreme authority or revered tradition. It is a piecemeal truth, conquered slowly and with great exertion, each small piece of which is all the more precious because it was so hard to attain. In the famous analogy of Otto Neurath, "We are like sailors who in the open sea must reconstruct their ship but are never able to start afresh from the bottom. Where a beam is taken away a new one must at once be put there, and for this the rest of the ship is used as support. In this way, by using the old beams and driftwood, the ship can be shaped entirely anew, but only by gradual reconstruction."[5] A ship of fools, drifting on the high seas, has often been taken as an image of humankind. In the present state of the ship, not all the beams have been tested, far from it, and its design is less than ideal. In the intellectual framework within which our decisions are made, not everything is scientific knowledge, and we are not able yet to fit all we

4. *The Open Society and Its Enemies*, chap. 24.
5. *Anti-Spengler* (Munich, G. D. W. Callwey, 1921).

know, think, and do into a coherent whole. We need it, but we don't have it. This is no reason to compromise with truth; as Maurice Merleau-Ponty puts it, "It cannot be expected from a philosopher that he goes beyond what he sees himself, nor that he gives directions he is not sure of. The eagerness of souls is no argument here; one does not serve the souls by half-truths and impostures."[6] The first duty of intellectuals is to tell the truth.

This is what Musil has to say about Ulrich, the hero of his great novel, himself a mathematician, which pretty much summarizes what I have tried to say:

> He hated men incapable of "suffering hunger in their souls for the love of truth," to use the words of Nietzsche, those who hold back, those who shun discussion and who seek comfort, who cuddle their souls with nursery tales and feed it religious, philosophical or imaginary sentiments, which are like bread buns dipped in warm milk, claiming that intelligence would feed it stones instead of bread. His opinion was that we find ourselves in this time committed to an expedition with the whole of the human race, that pride commands us to answer "not yet" to all useless questions and to lead one's life according to interim principles, while remaining aware of a goal that those coming after us will reach. The truth is that science has developed the idea of a raw and sober intellectual power which makes mankind's old metaphysical and moral representations simply unbearable, even though it can put in its place no more than a hope: that some day will come, in a long time, when a race of intellectual conquerors will settle in the valleys of spiritual abundance.

6. *Eloge de la philosophie: Leçon inaugurale faite au Collège de France* (Paris: Gallimard, 1953).

Finding the Small Diameter of a Convex Table

START FROM THE FIRST (LARGER) diameter of the table, and call it AB. The distance between A and B is the greatest possible between two points on that table. Take a point M_1 on the upper half of the table (above AB) and another point M_2 on the lower half (below AB). Denote by x the length of AM_1 and by y the length of BM_2.

As x and y vary, the segment M_1M_2 moves around. The smallest possible value for x is $x = 0$ (then M_1 lies at A), and the smallest possible value for y is $y = 0$ (then M_2 lies at B). Denote by d the distance between A and B. The greatest possible value for x is $x = d$ (then M_1 lies at B) and the greatest possible value for y is $y = d$ (then M_2 lies at A). Giving pairs of values (x, y) specifies a position for the segment M_1M_2:

$$(x = 0, y = 0) \text{ puts } M_1M_2 \text{ on } AB$$
$$(x = 0, y = 1) \text{ puts } M_1M_2 \text{ on } AA$$
$$(x = 1, y = 0) \text{ puts } M_1M_2 \text{ on } BB$$
$$(x = 1, y = 1) \text{ puts } M_1M_2 \text{ on } BA$$

The distance between A and B is easily figured out in these four cases. We find:

If $(x = 0, y = 0)$ then the distance between M_1 and M_2 is d
If $(x = 0, y = 1)$ then the distance between M_1 and M_2 is 0
If $(x = 1, y = 0)$ then the distance between M_1 and M_2 is 0
If $(x = 1, y = 1)$ then the distance between M_1 and M_2 is d

More generally, define $f(x, y)$ to be the distance between M_1 and M_2 when the length of AM_1 is x and the length of AM_2 is y. From the above, we have:

$$f(0,0) = d$$
$$f(0,1) = o$$
$$f(1,0) = o$$
$$f(1,1) = d$$

If we now graph the function $f(x, y)$ on the square defined by $o \le x \le d$ and $o \le y \le d$, we get two maxima at two opposite corners of the square, $(x = o, y = o)$ and $(x = 1, y = 1)$, and two minima at the two other corners. This means that the graph has two peaks at the two first corners, and by our general theorem on islands there must be a mountain pass somewhere on the square; let $(x = a, y = b)$ be its position. Setting M_1 at a distance a from A and M_2 at a distance b from B gives us the second diameter we were looking for.

Note that this second diameter neither maximizes nor minimizes the distance. Note also that there might be several passes, corresponding to several possibilities for the second diameter. This would be the case if, for instance, the billiard table has the shape of a rectangle with the four corners rounded off. There would then be, in fact, four diameters: two large ones, corresponding to peaks in the graph, and two smaller ones, corresponding to passes in the graph.

The Stationary Action Principle for General Systems

THE SIMPLEST SYSTEM in classical mechanics is the one-ball bil-liards. The motion of the ball is fully determined by its first impact on the cushion, that is, by a pair of numbers (x, y), where x gives the position of the impact on the edge and y the incoming angle. We shall refer to (x, y) as the initial state of the billiards.

For general systems, we need more variables. A solid body, for in-stance, spins as it moves. To specify its state at any time, we need ten variables: three to give the position of the center of mass, two to give the direction of the rotation axis, three to give the velocity of the center of mass, and two to give the speed of rotation and the displacement of the axis. The state of any system in classical mechanics can be described by the values of an even number of variables, say $(x_1, y_1, \ldots, x_N, y_N)$, the variables x_n identifying the position and the variables y_n the associated velocities. This amounts to prescribing a point in a space of $2N$ dimen-sions, called the *phase space* associated with the given system. The num-ber N, which is often called the *number of degrees of freedom*, can be quite large for complicated systems.

To describe the motion of a given system, we need one further ingre-dient, namely a function H on the phase space. The number $H(x_1, y_1, \ldots, x_N, y_N)$ is called the *energy* of the state $(x_1, y_1, \ldots, x_N, y_N)$. The phase space and the energy contain everything there is to know about the system; if we have figured them out, we can write down the equations of motion (but we may not be able to solve them). These are differential equations, so that, if we are given the state at time $t = 0$, the state at any later time t is fully determined.

The most striking fact about these equations is that they are conserva-tive, that is, that throughout the motion, the value of the energy stays pegged at its initial value. Say this initial value is h; then the trajectory is entirely contained in the set $H(x_1, y_1, \ldots, x_N, y_N) = h$, which is a hypersurface

S in the phase space. In other words, a trajectory starting on a certain energy level remains on that energy level. Some of these trajectories might be closed; they will correspond to periodic motions of the given system.

With any closed curve drawn on S, we can associate a number, which is the *action* along the given curve. Maupertuis' principle states that the closed curves which make the action stationary are trajectories of the system; that is, they satisfy the equations of motion. From then on to prove that such stationary curves actually exist is quite a step. This was finally achieved by Claude Viterbo in 1986; thanks to his result, we now know that periodic solutions exist under very general conditions.

Viterbo's method is similar in spirit (although very different technically) to the method we used to find the small diameter of a convex billiard table. Hofer and I then had the idea to define diameters for general systems. To do this, we consider the hypersurface S, defined by the equation $H(x_1, y_1, \ldots, x_N, y_N) = h$. By Viterbo's result, S carries at least one closed trajectory, and in fact usually carries infinitely many; we order them according to the values of the action along them. The smallest value will be called the "first diameter" of S. The next smallest will be the "second diameter," and so on.

These diameters enjoy remarkable properties. In the case of the billiards, there are only two: a large one, L, and a small one, l. Imagine now that we have two tables, the first one with diameters L_1 and l_1, and the second one with diameters L_2 and l_2. If the first table is contained in the second one, then it must be the case that both its diameters are smaller: $L_1 < L_2$ and $l_1 < l_2$. The same property holds for the "diameters" of more general systems.

This is the key to Gromov's uncertainty principle. Indeed, an uncertainty region around $(x_1, y_1, \ldots, x_N, y_N)$ is mathematically indistinguishable from an energy level: the boundary of the region is the hypersurface S. I still have not figured out a satisfactory physical interpretation for that identification. Bur the mathematics are clear. Uncertainty regions have "diameters," just like energy levels, and the correct use of these diameters quickly leads to a proof of the second uncertainty principle, which is different from the one Gromov originally gave, and which links it closely with the stationary action principle.

(BIBLIOGRAPHICAL NOTES)

I HAVE BEEN CAREFUL to acknowledge my sources in the text, and to check all the quotations. Some sources, however, have been more inspirational than others. For the first two chapters, I have made extensive use of the books of Alexandre Koyré, notably *Etudes Galiléennes* (Paris: Hermann, 1940), *Du monde clos á l'univers infini* (Paris: Gallimard, 1967), *Etudes d'histoire de la pensée scientifique* (Paris: Gallimard, 1973). For the second, the book by Paolo Rossi, *La nascita della scienza moderna in Europa* (Rome-Bari: Laterza, 1997) has been extremely useful. Of course, I have read Leibniz's *Monadology*, but if I had not come across the commented edition by Clotilde Calabi (Milan: Bruno Mondadori, 1995), I am afraid it would have remained a closed book to me.

From chapter 3 on, I have relied directly on the writings of the various actors, Fermat, Maupertuis, Voltaire, Euler, Lagrange, and on my expertise as a professional mathematician. I have also consulted historians of science, such as Ernst Mach and René Dugas, but I have allowed myself to disagree with them on occasion. In chapter 5, we enter completely new territory, since we are explaining mathematical discoveries which have happened during my lifetime, some of which I have contributed to, and there are of course no other references to them except in scholarly journals. Chapter 6 covers more familiar territory, chaos theory for instance, for which there are many references, including earlier books by myself, *Mathematics and the Unexpected* and *The Broken Dice*, both at University of Chicago Press (Chicago: 1990 and 1993).

Chapter 7 is the only one in the book where I cannot claim firsthand expertise. I am no biologist, and although I have looked up the original sources, notably Darwin, I have had to rely on others' work. The views I have found most congenial to my own thinking, and which are reflected here, are those of Stephen Jay Gould, as expressed, for instance, in *A Wonderful Life* (New York: W. W. Norton, 1989). I realize that almost twenty years have elapsed since that book was published, and that much progress has been made since then, notably in the understanding of the Burgess shale, but I have not updated my knowledge, in the belief that the core of the argument (that evolution cannot be interpreted as progress in any meaningful way) was unaffected.

With chapter 8, we go back to mathematics (optimization theory), and we introduce two authors whom I have studied for a long time, the Greek historian Thucydides and the Renaissance historian Guicciardini. There are many studies about Thucydides, the most recent (and most interesting) one being the book by Marshall Sahlins, *Apologies to Thucydides* (Chicago: University of Chicago Press, 2004), which appeared too late for me to take advantage of it. Much less has been written about Guicciardini, although it is my belief that he is no less a historian. Certainly he has been extremely useful to me in understanding the uses and limitations of political power. Chapter 9 introduces some basic economic concepts, such as efficiency (Pareto optimality), and problems, such as the difficulties of collective choice.

Of course, it is impossible in a couple of hundred pages to cover all the scientific and philosophical aspects of the problem raised by the title. What I have given here is by necessity a personal choice, colored by my training as a mathematician and as an economist, but it is a considered and careful one. Readers with different backgrounds, such as philosophers, anthropologists, or biologists, will probably find my account unbalanced. I hope some day we will reach a unified view of science and philosophy which will make a full account possible, perhaps along the lines drawn up by the Vienna Circle, but I fear this is as far away as grand unification is in physics. Meanwhile, readers who want to continue investigating these questions in an entertaining (yet profound) way, are directed to the science-fiction books of Stanislaw Lem. An excellent introduction to Lem, and a handbook for would-be creators, has been written by Bernd Gräfrath: *Es fällt nicht leicht, ein Gott zu sein* (Munich: Beck, 1998). Readers who want to learn more about Maupertuis' fascinating personality are directed to Mary Terrall's book, *The Man Who Flattened the Earth* (Chicago: University of Chicago Press, 2002).

Finally, I would like to recall the memory of Susan Abrams, who encouraged me to start this ten-year effort and kept supporting me until her untimely death.

accuracy, 3, 101, 104, 112, 127. *See also*
   precision
action
   in billiard, 110
   definition: by Euler, 66; by
      Leibniz, 63; by Maupertuis, 1,
      61–62, 66; by Hamilton, 69,
      196
   in quantum mechanics, 119
   See also least action principle
Agreement between Several Laws of
   Nature Which until Now
   Had Seemed Incompatible
   (Maupertuis), 61
Alembert, Jean Le Rond d', *Treatise on*
   Dynamics, 73
Alexandria, school of, 4–5
Analytical Mechanics (Lagrange), 68,
   79–82 passim, 73–74
Apology of Galileo (Campanella), 35–36
Archimedes, 5, 7, 11, 81, 150
Arcy, Chevalier d', 69
Arendt, Hannah, 183
Aristotle, 7; *Physics*, 5
Arrow impossibility theorem, 170
Athens and Sparta, 137–38
awareness. *See* purpose

baker's transformation, 125–27
belling the cat, 142–43
Bernoulli, Jacob, 18–20, 80, 154

Bernoulli, Johann, 18–20, 80, 154
best of all possible worlds, 1
   according to Leibniz, 36–43; to
      Maupertuis, 63, 129
   and evolution, 130
   See also criterion
billiard
   action in, 110
   chaos in (*see under* chaos)
   circular, 89–92, 105
   convex, 89, 104–5, 109–11, 193
   elliptic, 92–94, 100, 105
   general (nonelliptic), 97–101
      passim, 110, 113–15
   motion in (*see under* motion)
   periodic motion in (*see under*
      periodic trajectory)
   with several balls, 124–25
   trajectory in (*see under* trajectory)
Bohr, Niels, 56–57
Borda procedure, 170, 173
Borel, Pierre, *New Discourse Proving*
   That Celestial Bodies Are In-
   habited Earths, 35
Borges, Jorge Luis, 39
brachistochrone, 17–20, 154
Brecht, Bertolt, *Life of Galileo*,
   25–26, 33

calculus of variations, 20, 66, 81,
   151–55 passim

Campanella, Tommaso, *Apology of Galileo*, 35–36
Candide (Voltaire), 1, 63
Cartesians, 54, 57–58
causal chain, 85–88, 91, 94–97 passim
causality, 85, 95
cause
    effects proportionate to, 48, 87, 94
    efficient vs. final, 28, 31–32
    multiple, 86–87
    none, 9
chance, 63, 85, 138, 144. *See also* randomness
chaos, chaotic
    in baker's transformation, 125–27
    in billiards, 95–100, 110, 124–25
    in celestial mechanics, 84, 103
Chomsky, Noam, 188–89
Churchill, Winston, 168
Cleopatra, 86–87
clepsydra, 15
Clerselier, 48, 75, 118, 121
clock. *See under* time
closed. *See* periodic
Condorcet, Marie Jean Caritat, marquis de, *Discussion on the Application of Analysis to the Probability of Decisions Taken by Majority Voting*, 169
Condorcet paradox, 170
conics, 5
convex, table. *See* billiard, convex
coordinate
    horizontal, 90
    vertical, 90
criterion
    for best of all possible worlds, 41, 129, 167
    in economics, 172–74, 179
    in evolution, 130–31
    in optimization, 129, 151, 154–57
    for truth, 24

Cureau de la Chambre, 50, 53; *On Light*, 50

Darwin, Charles
    and descent with modification, 131, 133
    Origin of Species, The, 133
decision-making
    collective, 169–70
    individual, 138
Descartes, René
    and experiment, 48
    on light, 49–50, 53–54, 61–62, 118
    on science, 24–25, 31–33, 47–48
    and trial of Galileo, 46
    works: *Discourse on the Method, A*, 46–47, 185; *Geometry*, 46–47; *Meteors*, 46; *Optics*, 46; *Treatise of the World, or Of Light*, 46–48
descent with modification, 131–36 passim, 143. *See also* struggle for life, survival of the fittest
determinism, 55–56, 75
Dialogue on the Two Greatest Systems of the World, the Ptolemaic and the Copernican, (Galileo), 46
diameter, 92, 105–6, 109, 193
Diderot, Denis, *Encyclopédie*, 151
Discourse on the Method to Orient One's Reason and to Seek Truth in Sciences, A (Descartes), 46–47, 185
Discourses and Mathematical Proofs Concerning Two New Sciences (Galileo), 3, 17
Discovery of a New World (Wilkins), 35
Discussion on the Application of Analysis to the Probability of Decisions Taken by Majority Voting (Condorcet), 169
Discussions on the Plurality of Worlds (Fontenelle), 35

*Dissertation on the Least Action
   Principle* (Euler), 67
dualism, 31, 36

Easter Island, 146, 148
efficiency, 152, 171–72
Einstein
   God does not play dice, 55, 121
   theory is good, 17
   See also relativity
Elements of the Calculus of Variations
   (Euler), 81
Encyclopédie (Diderot), 151
energy, 195–96
engineers
   humans as, 150
   scientists as, 33, 42, 151–52
   use optimization theory, 158
entropy, 166
environment, 1, 132, 147–48, 158
equations
   of Euler-Lagrange (*see* Euler-
      Lagrange equations)
   of motion, 25, 27, 79, 81, 110, 154–55
   solving, 25, 27, 80–84 passim, 101
equilibrium
   in biology, 133–34
   in game theory and economics,
      140–41, 177, 179
   in politics, 139
   in physics, 7–8, 16, 81, 123
Essay in Cosmology (Maupertuis), 62
Euclid, 10
Euler
   and calculus of variations,
      80–83, 154
   and least action principle, 66–67,
      70–74, 81
   works: *Dissertation on the Least
      Action Principle*, 67; *Elements of
      the Calculus of Variations*, 81;
      *Method for Finding Curves
      Which Are Maximizing or
      Minimizing, A,* 62, 73, 80;

On the Motion of Projectiles, 74;
   *Theory of Motion of Solid or
   Rigid Bodies,* 82
Euler-Lagrange equations, 67, 83,
   154–55
Euler's formula, 106, 108
evolution, 60, 130

falling bodies, 15–17, 22–23
Fermat, Pierre de
   on light, 50–54
   on science, 54–57
Feynman, Richard, 119
Feynman's principle, 120
filtering, 158
Finer, S. E., *History of Government
   from the Earliest Times,* 159
fitness, 40, 130–33
foliation, 94
Fontenelle, Bernard de, *Discussions
   on the Plurality of Worlds,* 35
forecasting, 94. *See also* prediction
French Revolution, 33, 159

Galileo
   and experiments, 15–17, 22–23
   and fieldglass, 34, 45, 151
   on mathematics and science, 24–26
   and pendulum, 3–4, 8–11, 20–21,
      150
   on space and time, 5, 7, 11–14, 23,
      27–29
   trial of, 9, 44–46
   works: *Dialogue on the Two Great-
      est Systems of the World, the
      Ptolemaic and the Copernican,*
      46; *Discourses and Mathemati-
      cal Proofs Concerning Two New
      Sciences,* 3, 17
game theory, 139, 141
geometry
   and mathematics, 25, 30, 47
   and mechanics, 75, 109, 111
   and physics, 5, 9, 11, 27, 57

Geometry (Descartes), 46–47
global warming, 146–47
God
    no need for, 27–28, 164
    rests after Creation, 31
    See also purpose: of God in
        creation; reason: binds even
        God
good, common, 166–69, 179–80
Gould, Stephen Jay, *Wonderful Life,*
    A, 135, 197
Gracian, Balthazar, 163–64
gravitation, law of, 29–30, 34, 64,
    103, 182
Gromov, Mikhail, 111–16, 183, 196
Guicciardini, Francesco
    History of Italy, 136, 138–39,
        160–61
    personal experience, 160–62,
        164–65
    Ricordi, 138, 162

h. See Planck's constant
Hamilton, William Rowan, 66,
    68–69, 79
happiness, 184
Harrison, John, 14, 22, 151
Hero of Alexandria, 50–51
Hilbert's problems, 155
History of Government from the Earliest
    Times (Finer), 159
History of Italy, 136, 160–61
History of Mechanics (Mach), 56, 67,
    75–77
History of the Peloponnesian War
    (Thucydides), 136, 160–64
    passim, 198
History of the Roulette (Pascal), 18
Horologium Oscillatorum (Huygens),
    21
hours, length of, 4
Hutcheson, Francis, 169
Huygens, Christiaan
    and clocks, 21–22, 150

criticizes Descartes, 48
Horologium Oscillatorum, 21
and isochronous pendulum, 19,
    80
and light, 23, 118–19

implementation, 174, 179
information
    asymmetry of, 174
    cannot be created or transferred,
        113–15
    lost in chaos, 125
    about nature, 46
integrable system, 80–87 passim,
    95, 100–101, 154–55
interest, common. *See* good,
    common
inverse square law. *See* gravitation,
    law of
isochrony, 3–4, 8–9, 16–20

Jacobi, Carl Gustav 62, 79, 81–82,
    97; *Lectures on Dynamics*, 66,
    71–72, 74–75

Kant, Immanuel, 188
Kepler, 26
Kepler's laws, 27, 64, 102–3
kinship, 147
Kovalevskaya, Sofia, 83
Koyré, Alexandre, 197

Lagrange, Joseph Louis, 66–68,
    70–74, 79–83 passim, 97, 154;
    Analytical Mechanics, 68,
    79–82 passim, 73–74
Laplace, Pierre Simon de, 27–28,
    164
Laws of Motion and Rest Deduced from
    a Metaphysical Principle
    (Maupertuis), 62
laws of nature, 180
leaf (of foliation), 94
least action principle, 1, 44

Euler and, 66–67, 70–74, 81;
    Leibniz and, 63–64; Mach
    and, 56, 67; Maupertuis and,
    56–57, 61–63, 70–76 passim,
    152, 182–83
    mistrust of, 68, 73–75, 79
    wrong as stated, 62, 69–75, 129
    See also under best of all possible
    worlds; Hamilton; Jacobi;
    Lagrange; Leibniz; purpose;
    stationary action principle
Lectures on Dynamics (Jacobi), 66,
    71–72, 74–75
Leeuwenhoek, Antony van, 42
Leibniz, Gottfried, 36
    as naturalist, 42–43
    and possible worlds, 36–43
    and predestination, 38
    works: *Monadology*, 40–41, 197,
    Theodicy, 38
    See also under best of all possible
    worlds; possible worlds
Lem, Stanislaw, 197
    Solaris, 187
length, unit of, 10
Leonardo da Vinci, 151
Lévi-Strauss, Claude, 187
Lichtenberg, Georg, 188
Life of Galileo (Brecht), 25–26
light
    consists of particles, 49, 61–62;
    of waves, 23, 183
    speed of, 12–13, 49–50, 53–54,
    61–62
    takes quickest path, 51–54, 61
    takes shortest path, 50, 61, 69
    See also reflection; refraction
Light, On (Cureau de la Chambre), 50
Liouville, Joseph, 112–13
longitudes, measuring, 14
Lucretia, rape of, 38–39

Mach, Ernst
    History of Mechanics, 56, 67, 75–77
    on science, 67, 76–77
    See also least action principle
Machiavelli, Niccolò, 163–64
majority voting, 169
Malebranche, Nicolas, 31, 33
Malgas and Marcus islands, 132–34,
    141
Man without Qualities, The (Musil),
    189, 191
marine chronometer, Harrison's, 14,
    22, 151
Mark Anthony, 86
mathematics, 6, 23, 24–25, 30, 47–48
Maupertuis, Pierre Louis Moreau de,
    1, 35
    Northern expedition 58–59
    quarrel with Voltaire 60, 63–66
    works: *Agreement between Several
    Laws of Nature Which until Now
    Had Seemed Incompatible*, 61;
    Essay in Cosmology, 62; *Laws of
    Motion and Rest Deduced from a
    Metaphysical Principle*, 62;
    Physical Venus, The, 60; *White
    Negro, The*, 60
    See also least action principle, best
    of all possible worlds
maximizing, 129
Measuring the Circle (Archimedes), 11
mechanics
    analytical approach to, 25, 68, 79
    classical, 20, 25, 84–85, 111–16
    passim, 122, 154–56
    Feynman's interpretation of,
    120–22
    geometrical approach to, 68, 111
Meno (Plato), 6
Merleau-Ponty, Maurice, 191
Mersenne, Marin, 16, 22
Meteors (Descartes), 46
*Method for Finding Curves Which Are
    Maximizing or Minimizing, A*
    (Euler), 80
minimizing, 129

Mittag-Leffler, Gösta, 102–3
model, mathematical, 53, 57, 76
Monadology (Leibniz), 197
Morgenstern, Oskar, *Theory of Games and Economic Behavior*, 166
motion
    in Aristotle's physics, 5–7
    in billiards, 88–89
    equations of (*see under* equations)
    possible, 66, 68, 75
    on a sphere, 72
    transitory, 7–8, 15
    types of: circular, 25–26, 33; elliptic, 25–28, 33; linear, 26, 66; uniform, 25–26, 66
    See also pendulum, rigid body, trajectory
Motion of Projectiles, On the (Euler), 74
mountain pass theorem, 106
Musil, Robert, *Man without Qualities, The*, 189, 191

n. See refraction index
Nash, John, 139
nature, 145–50 passim
Neumann, John von, 139; *Theory of Games and Economic Behavior*, 166
Neurath, Otto, 190
New Discourse Proving That Celestial Bodies Are Inhabited Earths (Borel), 35
New Methods of Celestial Mechanics, The (Poincaré), 102, 109
Newton, Isaac
    cannot solve motion of Moon, 103
    and Cartesians, 57–58
    and experiment, 28–29
    finds shape of least resistance, 152–54
    on light, 50, 118
    Philosophiae Naturalis Principia Mathematica, 28, 30, 58, 64, 152, 154

on science, 28–30, 36
solves two-body problem, 27, 64, 80, 102
See also gravitation
new worlds, discovery of, 34–36, 182
nonintegrable system, 87, 95–96, 100–101, 103–4
nuclear war, 145–46, 148

Ockham's razor, 27–29
Oedipus, 38, 147–48
Open Society and Its Enemies, The (Popper), 186–87, 189–90
operations research, 156
Optics (Descartes), 46, 49
optimal shape
    of least resistance, 152–54
    of least time (*see* brachistochrone)
optimization, optimizing, optimum, 1
    in economics, 172–73
    in engineering, 152, 156–58
    and evolution, 129–33 passim
    and game theory, 140
orbit. *See* trajectory
Origin of Species, The (Darwin), 133

π, 11
Pandora's box, 123–25, 127–28
Pareto criterion, 171–72
Pascal, Blaise
    History of the Roulette, 18
    Thoughts, 163–64, 178
pendulum
    construction, 3, 21–22
    ideal or Galilean, 3–4, 8–11, 17, 23
    as an integrable system, 87
    truly isochronous, 18–19
    used to measure time, 4, 8, 10–11, 150
    See also isochrony; period; time
perfect, perfection
    according to Darwin, 133

according to Leibniz, 39–41
according to Plato, 6
in pendulum motion, 7
period
  of pendulum, 3, 9, 10: dependence
    on amplitude, 8, 9, 16–17;
    dependence on latitude, 22;
    dependence on length, 3, 8, 17;
    dependence on temperature, 22
  of periodic trajectory, 103–4
  of planets, 103
periodic trajectory
  in billiards, 90, 104–6, 109–11
  in general systems, 103, 110, 196
Phaedo (Plato), 32
phase space, 68, 75, 195
*Philosophiae Naturalis Principia
  Mathematica* (Newton), 28,
  30, 58, 64, 152, 154
Phragmen, Edvard, 102
Physical Venus, The (Maupertuis), 60
Physics (Aristotle), 6
Planck's constant, 120–21
Plato: *Meno*, 6; *Phaedo*, 32
Poincaré, Henri
  and mechanics, 25, 79, 84, 97,
    109, 111
  theorem on time reversibility,
    123–24
  and three-body problem, 102–4
  works: *New Methods of Celestial
    Mechanics, The*, 109; *Science
    and Hypothesis*, 77; *Value of
    Science, The*, 77–78
Pontryagin, Lev, 157
Pontryagin's principle, 157
Popper, Karl, 29
  Open Society and Its Enemies, The,
    186–87, 189–90
possible worlds, 34, 129, 148
  according to Descartes, 47–48
  according to Leibniz, 36–43
  See also universe: conceivable
power, 140, 164

precision
  of instruments, 127
  in locating a position, 84, 95
  in measuring time, 15–16, 20,
    21–23
  physical vs. mathematical, 11
  See also accuracy
prediction, 87–88, 91, 95, 125–27.
  See also forecasting
preferences 167–68, 170–71
principle
  of inertia, 56
  of minimum time, 56–57
  physical, 53, 67, 73
  of stationary path, 75
  See also least action principle;
    Pontryagin's principle;
    uncertainty principle:
    Gromov's
probability
  and chaos, 127–28
  in quantum physics, 57,
    120–21
Ptolemy, 26
purpose
  of God in creation, 1, 30–33, 57,
    180
  not present in nature, 55–56,
    145, 166, 182
  revealed in least action principle,
    62–63, 65, 71–72, 78

randomness
  and chaos, 125–26
  driving evolution, 134–35
  driving history, 136
  in quantum physics, 55, 119,
    121
  See also chance
ranking, 129, 167–70 passim
rational decision, 145, 186, 190
rationalism, 186–87
Rawls, John, 173–74
  Theory of Justice, A, 173

reason
    binds even God, 33–34, 36–37, 47
    and human nature, 25–26,
        186–90
redistribution, 173, 179
reflection
    on plane, 50–51
    on sphere, 69
refraction, 49–54 passim, 57, 61–62,
    152
refraction index, 49, 52–54
relativity, Einstein's theory of
    general, 27–28, 182
    special, 13
Renaissance, 33–34, 151, 165, 180
Ricordi (Guicciardini) 138, 162
rigid body, 67, 82–84
Rorty, Richard, 159
Rossi, Paolo, 197
roundoff error, 104
Rousseau, Jean-Jacques, 167–68
roulette, 18–22

Sahlins, Marshall, 197
science
    role and scope of, 24, 30, 75–78,
        185
    simplicity in, 26–29, 41, 76, 78
    and technology, 20–21, 33, 77,
        150–51
    See also under Descartes; Galileo;
        Mach; Newton; Poincaré
Science and Hypothesis (Poincaré), 77
scientific method, 185–86, 188
selection, natural, 133
simplicity. *See under* science
simultaneity, 12–13
Snell's law, 49
social conventions, 140, 163, 178
social order, can be changed, 159,
    164–65
social value, 173
society, organization of
Solaris (Lem), 187

solution
    general, 101, 154–55
    particular, 154–55
Sparta and Athens, 137–38
stationary action principle
    results from randomness on
        lower scale, 117–22 passim
    statement, 68–69
    used to find periodic trajectories,
        103–4, 109, 111, 182–83,
        195–96
    useless in integrable systems, 100
stationary path, 69–70, 75, 119
stationary point, 69–72, 108–9,
    117–18, 151–52
stochastic control, 158
*Story of Doctor Akakia and the Native
    of Saint-Malo* (Voltaire), 65
strategic behavior, 140, 174
struggle for life, 132–35 passim. *See
    also* descent with modification
sundial, 15
survival of the fittest, 130, 132

technology. *See under* science
Theodicy (Leibniz), 38
theory, 7, 29
*Theory of Games and Economic
    Behavior* (Neumann and
    Morgenstern), 166
Theory of Justice, A (Rawls), 173
*Theory of Motion of Solid or Rigid
    Bodies* (Euler), 82
thermodynamics, 122
Thoughts (Pascal), 163–64, 178
three-body problem, 102–4
Thucydides, 160–62, 164–65;
    *History of the Peloponnesian
    War*, 136–38, 160–64 passim,
    198
time
    infinite, 9
    instruments for measuring, 8–9,
        14–15, 20–23, 150

irreversibility of, 122–24, 128
unit of, 10, 23
universal, 3, 13–14, 27
See also under pendulum
time's arrow, 122–25
top, 83
trajectory
in billiards, 90–94, 96, 100
in collisions, 55
closed (*see* periodic trajectory)
of motion, 82, 112–14, 122, 155
of planets, 26–28, 30
of projectiles, 6
See also motion
Treatise on Dynamics (d'Alembert),
73
Treatise of the World, or Of Light
(Descartes), 46–48, 54
trust, 140–41
two-body problem, 80, 102. *See also*
Kepler; Newton

uncertainty, 112–15, 147, 157
uncertainty principle
Gromov's, 111–16, 183, 196
Heisenberg's, 111
universe
conceivable, 135
infinite, 10, 12, 34, 42–43
unstable, 85, 110
utilitarian criterion, 173

Value of Science, The (Poincaré),
77–78
Venetian senate, 138–39
Viterbo, Claude, 196
Voltaire, François-Marie Arouet
Newtonian, 58
quarrel with Maupertuis, 1, 43,
60, 63–68
works: *Candide*, 1, 63, 65–66;
*Letters concerning the English
Nation*, 58; *Micromegas*, 35,
63; *Story of Doctor Akakia
and the Native of Saint-Malo*, 65

watch. *See under* time
waves
difference with particles, 119
of light, 23
on water, 118
Weber, Max, 142
welfare, 169. *See also* good, common
White Negro, The (Maupertuis), 60
Wigner, Eugene, 24
Wilkins, John, *Discovery of a New
World*, 35
Wittgenstein, Ludwig, 78
Wonderful Life, A (Gould), 135, 197
world
as a machine, 30–33, 42
new, 48
See also possible worlds